BONDING
and
STRUCTURE

A REVIEW OF Fundamental Chemistry

EMIL J. MARGOLIS

THE CITY COLLEGE OF
THE CITY UNIVERSITY OF NEW YORK

APPLETON
CENTURY
CROFTS

NEW YORK DIVISION OF
MEREDITH
CORPORATION

Copyright © 1968 by
MEREDITH CORPORATION

All rights reserved

This book, or parts thereof, must not be used or reproduced in any manner without written permission. For information address the publisher, Appleton-Century-Crofts, Division of Meredith Corporation, 440 Park Avenue South, New York, N.Y. 10016.

688-2

Library of Congress Card Number: 68-12785

PRINTED IN THE UNITED STATES OF AMERICA
E 59789

PREFACE

MODERN *orbital theory* ENDEAVORS TO PROVIDE DETAILS OF THE conceptual nature and degree of the forces of attraction and repulsion—that is, the extent of their *bondings*—that are mutually exerted by atoms on one another. Rational interpretations of bondings which are predicated, in the main, upon experimentally observed architectural forms or *structures* of chemical species, assist both in the predicting of new chemical phenomena and in the logical organization and correlation of those already familiar. We appropriately concern ourselves, therefore, in this volume with the orientations of electrons postulated by various current orbital theories of bonding and structure: *valence bond* (VB); the *electrostatic crystal field* (CF), and the *molecular orbital* (MO) proper. When appropriately related to other aspects of the subject, each reveals its relative merits in explaining chemical bonding and in interpreting chemical properties.

In its breadth of development and attention to detail, this book seeks to compensate for the frequently unfulfilled obligations with respect to chemical bonding and structure imposed upon the comprehensive general chemistry textbook. By no means is this observation intended as an unsympathetic criticism. Indeed, the many diverse and frequently unrelated topics conventionally demanded for the first-year chemistry curriculum make it virtually impossible for any comprehensive text to delineate all significant facets of the subject. Where the attempted treatment proves superficial, ideas already sufficiently complex to the average beginner become aggravatingly complicated. This book prudently maintains itself, consequently, upon an educational level that does not presuppose or require completion of the basic introductory year.

It will thus serve to advantage as a supplement to a course in General Chemistry; and as a handy reference for Qualitative Analysis. It is in this latter area that the theoretical principles of bonding in general are put to practical experience, especially in the complexing of ions.

It will also be helpful as a basic text for college courses planned for advanced placement of qualified high school students; as a syllabus for inorganic chemistry for science teachers and teachers-in-training in Schools of Education; and as a book for intensive review before examinations on the integrated comprehensive topics covered.

The descriptive and computative exercises that have been provided are designed not only to test the student's understanding of the actual content of each chapter, but, also in significant numbers, to introduce and interpret new factual material. Opportunity is thus extended to acquire additional information even while demonstrating an ability to apply the principles already developed. In support of this approach, all exercises have been fully answered in Appendix A.

E. J. M

CONTENTS

CHAPTER

ONE	Atomic Properties	1
TWO	Properties of Chemical Bond	29
THREE	Complexing Transitional Metals	79
FOUR	Aspects of MO Theory	125

Appendix A: Answers to Exercises	149
Appendix B: Energy Units, Fundamental Constants and Reference Symbols	161
Appendix C: International Atomic Weights	161
Appendix D: Periodicity of Properties of the Elements	165

Index   171

BONDING
and
STRUCTURE

CHAPTER

ONE

ATOMIC PROPERTIES

IN THIS CHAPTER WE INVESTIGATE THE ATOMIC PROPERTIES OF *ionization potential*, *electron affinity*, and *electronegativity*. These concepts eventually constitute the contributory components of an over-all visualization and justification of the chemical bond, with special emphasis upon *polar covalent* and upon *ionic* (electrovalent) bonds.

CAUSES OF CHEMICAL BONDING

If we are to conjecture any formulated substance — compounded or elementary, whether neutral and net-uncharged or electrically charged as an ionentity — as a discrete aggregation or association of atoms or groups of atoms, then we must logically inquire into the nature of the forces that permit such atomic intimacies. In essence, then, we ponder the chemical bond. As our most elementary concepts of matter firmly establish its inherent electrical make-up, all chemical interactions between and among atoms must be viewed as completely electrical in origin and effect. We may conclude, consequently, that each of two isolated but, potentially, chemically active atoms approaching one another seeks for its particular valence shell the full complement of electrons that endows it with the *greater stability* characterized and attributed to the inert noble gas of its pertinent sequence or series of periodic classification.

The electron clouds that constitute the *orbitals* of each atom (loci or orientations in space, which *probabilize* the presence of one or of a pair of electrons) mutually penetrate each other in response to the attractional pull of each positively-charged atomic nucleus for all electrons in its vicinity — those of the approaching atom as well as those of its own. As each atomic nucleus draws the other's associated electrons closer and closer, the resultant progressive overlapping of the electron clouds of the atomic orbitals (AO) reach a point of maximum mutual penetration. This is established by the balanced equilibrium oppositions of progressively increasing electron-electron repulsions which occur as the distance between the atomic nuclei progressively diminishes. In this manner, two atoms mutually seeking the electrical make-up that may provide each with the configuration isoelectronic (the same number of electrons) with that of the noble gas of its periodic sequence, form a molecular orbital (MO) which is visualized as merely the coalescence-via-overlap of hitherto separate atomic orbitals.

As it must be presumed that no previously separated or isolated chemical species associates with another unless in so doing each fulfills its objective of greater stability, it is entirely plausible that the greatly enhanced stability enjoyed by the electrons in the molecular orbital accounts for the cause of the chemical bond. It is to be stressed that the sharing-in-common by atoms in the molecular orbital of its total electrical atmosphere in no way precludes a favored displacement of electron density to one or the other. In fact, the ionic or electrovalent bond must be regarded as the extreme case of such favored displacement. Any characterization of a chemical bond as ionic rather than nonionic (covalent) becomes a matter not of intrinsic or inherent nature but, rather, one of degree; and the component factors involved in their categorical differentiation are fully related and interdependent.

Our theoretical visualization of atomic orbital overlap imposes the necessary recognition of an important adjunct; namely, that the electrons in the now-formed molecular orbital have permanently lost their former identity with the original isolated atoms whence they derived.

All electrons, and the nuclei of the hitherto separate atoms as well, are now no longer to be considered as belonging to atomic orbitals; as a result of their mutual interaction, they are now fully and exclusively molecular-orbital components. Nonetheless, as we have already indicated, it does not follow that the electrons will necessarily divide their time of kinetic motion in residually equal fashion between the chemically bonded atoms in the molecule. The situation wherein the electrons spend their time equally with each nucleus of the molecular orbital would have to occur necessarily only in the very simplest cases, such as two chemically identical atoms bonding to form a diatomic molecule. In such instances, there would

ATOMIC PROPERTIES

certainly be no reason to suppose that either of the two identical atoms would have any greater capacity to draw the electrons more closely than the other. Logically, the density of the electron clouds upon both such atoms would be precisely similar, although the proportionate distribution of the electron clouds in the extranuclear regions between the two nuclei would be much greater than in the peripheral regions. The greater concentration of negative charge that is concomitant with contraction of molecular-orbital shape along the internuclear axis also reflects the increase in the theoretical probability of locating the bonding electrons. This may be appraised from the empirical representations made of the hydrogen atom, in Figure 1.1.

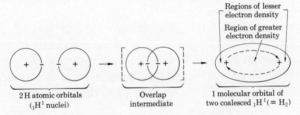

FIG. 1.1 Formation of a molecule of hydrogen [2H → H_2].

When the atoms in a chemical bond are different from each other their respective capacities to draw additional electrons to themselves are likewise different; hence, one atom in the molecule may be favored with a significantly greater electron density than the other. We logically conclude consequently, that when bonding electron pairs are not shared equally a residual increment of negative charge characterizes that atom in the molecule which has the greater attraction for electrons. In parallel, a residual increment of positive charge then characterizes the atom of lesser attraction for electrons. It must be understood that the entire molecule is itself completely neutral electrically when considered as a net entity. All that has occurred is merely a geographical shift or displacement of the bonding electrons from what otherwise would have been the point of equal distribution between the affected atoms.

The important point to be made here in qualitative fashion (pending later quantitative delineation) is that the extent to which the bond between the atoms can be maintained is determined both by the quantity of increment of charge on each of the separate molecular sites — electronegative and electropositive — and by the distance between these sites. Herein, then, is the basis for the completely ionic or electrovalent bond on the one extreme and the purely covalent bond on the other; and the inevitable "in-between" innumerable bondings of both partial ionic and partial covalent character.

The presumed 100% ionic bond is the idealized extreme of virtually complete displacement of the bonding electron pair to one atom at the expense of the other; that is, the removal of the bonding electron pair by one atom to an infinite distance from the other. The 100% covalent bond, on the other hand, would be attributable to an exactly equal distribution between the atoms of the charge density of the bonding electron pair. As we have seen, this would be the case of a diatomic molecule comprised of chemically identical atomic nuclei (e.g., the hydrogen molecule). The highly variable "in-between" partial-ionic, partial-covalent bond that categorizes nearly all of the vast numbers of known chemical substances reflects the many differences of chemical bonding strength that makes for differences in chemical and physical properties. Apropos of the critical theoretical import of bonding strength, it must be borne in mind that *all* chemical interaction necessarily involves both the breaking of old bonds and the forming of new ones.

In the foregoing, we have given thought to the over-all qualitative significance of electron attractions insofar as such attractions differentiate definitively the various types of bonds. We now delineate the attributes of the atoms presumed to constitute the causative and contributory components of the over-all attraction referred to as the *chemical bond*.

IONIZATION POTENTIAL; IONIZATION ENERGY

Both *ionization potential* and *ionization energy* mean the work necessary to pull from the net-neutral *isolated* gaseous atom its most loosely held electron and leave, consequently, a positively-charged monovalent gaseous ion. This most weakly held electron is always the one of greatest energy (with reference to those remaining), and separating it must be regarded as complete removal to a distance of infinity from the positively charged nucleus with which it was originally associated. This characterizes all ionization processes. The formation of a monovalent cation represents the original intent of the concept of ionization potential or ionization energy and it is rigidly to be accepted in this meaning unless otherwise modified.

With interests extended to the removal of additional electrons from the monovalent cation as well, it becomes necessary to employ the modifying alternatives of successive ionization potentials or energies, conforming, respectively, to the following sequence in removing electrons from the generalized gaseous atom, X:

1st ionization potential or ionization energy: $X^0_{(g)} \rightarrow X^+_{(g)} + 1e^-$
2d ionization potential or ionization energy: $X^+_{(g)} \rightarrow X^{2+}_{(g)} + 1e^-$
3d ionization potential or ionization energy: $X^{2+}_{(g)} \rightarrow X^{3+}_{(g)} + 1e^-$
nth ionization potential or ionization energy: $X^{(n-1)+}_{(g)} \rightarrow X^{n+} + 1e^-$

As would be logically anticipated from the progressive build-up of cationic charge in such a sequence, the parallel energy requirements for the pertinent electron-removals increases considerably with each progressive step of the sequence. It must not be thought, however, that the energy required to remove two electrons is twice that necessary to remove one. Nothing could be farther from reality. The differences in ionization potential or energy for such sequential electron-removals are *disproportionate*, and in most instances, enormously so.

That this is not an illogical situation can be readily interpreted from the following:

1. Relative to the *atomic number* of the element — that is, the atom's *actual* positive charge representative of its precise number of nuclear protons — the extent of such positive charge is a highly important determinant of the degree of coulombic attraction to which the negatively charged electron responds.

2. Relative to the *dimensions* of the atom or ion of interest, the radius of the atomic or ionic species is the specific factor to be considered inasmuch as we may acceptably visualize the species as roughly spherical in shape — although they are highly deformable electron clouds.

3. Relative to the *orbital level* from whence the particular valence-shell electron responding to ionization potential or ionization energy is removed, inasmuch as the distance from the nucleus of an electron in a particular valence shell quite generally increases in the progressive order, $s < p < d < f$ it follows that in any given valence-shell the s electron would be, ordinarily, the one least readily removed from the electrostatic attraction of the nucleus.

4. Relative to the extent to which electrons in *innermost* shells screen or shield those electrons in the *outermost* electron shell that are subjected to ionization, any increase in the number of electrons between the nucleus and the valence shell operates to increase the *shielding effect*. This results from the increase in the extent to which (a) the added inner electrons neutralize the positive nuclear electrification attracting electrons of negative valence; and (b) the electron(s) of the valence shell repulse the now-greater numbers of electrons present in the innermost shells.

A useful term that qualitatively equates the shielding or screening effect to the actual nuclear charge is *effective nuclear charge*. Logically enough, this latter term is the actual nuclear charge *less* the screening effect. It would normally follow that the greater the screening effect the smaller would be both the effective nuclear charge and the ionization potential or ionization energy required to rip an electron away from its valence shell.

The screening effect also manifests itself in orderly contractions in the radii of atoms as we move from left to right in a given sequence (or period) of the representative elements of periodic classification (i.e. main groups). These atomic radii are calculated on the basis of atomic numbers (actual nuclear charges) and external electron distributions. As positive charge is added to each nucleus in orderly sequence, corresponding additions of electrons, likewise, must be made to the outer valence shell. Because screening effects of these outermost electrons are trifling, however, compared to those in inner shells, they offer virtually no significant compensations of negative charge to oppose the progressively increasing pull of the nucleus on the fixed and unchanged numbers of inner-shell electrons. With effective nuclear charge thus becoming progressively larger in proceeding from left to right in any representative period of elemental classification, atomic and ionic radii progressively diminish in the same direction.

With respect to a periodic group (or family) of elements, atomic radii of the representative elements quite generally increase from top downwards in any classification. This parallels an increase in atomic number of each succeeding member of the chemical family; and hence such radial-increase is in direct opposition to the contraction to be expected, were increase of nuclear charge the sole consideration. Clearly, however, as a net effect, the steadily increasing contraction that results from the increasing nuclear charge as we proceed downwards in a periodic family group is more than offset by the opposing enlargement of the size of the atom as more electron shells are added. In the representative elements, this latter effect of screening or shielding of outermost electrons by the added inner shells is the more important effect.

The changes in atomic sizes of the *transition* elements follow much the same pattern as those of the representative elements — although the numerical differences in the sizes of radius between succeeding members of the respective periods and of family groups are, on the whole, of smaller degree and not quite so regular as those of the representative elements. The prime characteristic of the transition elements, which serves to distinguish them from the representative elements, is their belated filling of inner electronic shells while essentially maintaining the status of the electrons in their outer valence shells.

MEASUREMENTS OF IONIZATION POTENTIAL (OR IONIZATION ENERGY)

Ionization potential is, perhaps, the most fundamental of all attributes of an atom that are susceptible of measurement by experiment. This property may be determined readily from studies of the spectrum of the gaseous species or from the latter's behavior when under low pressure in a gas-discharge tube. In the latter instance, when the voltage is steadily gradually increased the hitherto continuing absence of any significant corresponding flow of current across the two plates of the electrical field is very abruptly terminated with a well-defined and greatly increased voltage. From this *critical potential* obtained, the energy required for ionization may be calculated. When energy is expressed in the units of electrical energy — *electron volts*— we properly speak of *ionization potential*; when expressed in the units of chemical energy — *calories* or *kilocalories* — the proper terminology is *ionization energy*.

When ionization energies are to be predicted or calculated from spectral studies of the element, the quantum theories of Planck and Einstein become indispensable tools. These, in their postulations that electromagnetic radiations have the properties of both corpuscles and waves, have led to the eventual derivations,

$$E = h\nu$$

and

$$\nu = c/\lambda$$

wherein

E = energy quantum of the electron,

h = Planck constant — a proportionality constant having a universally fundamental value of 6.63×10^{-27} erg·sec,

ν = frequency of the radiation (conceived as a stream of photons of light),

c = velocity of light (3.0×10^{10} cm/sec),

λ = wavelength of light, in centimeters, corresponding to the wavelength of the radiations established for the limits of energy transmissions in spectrum of the particular element.

For purposes of illustration we utilize the preceding relationships to evaluate the ionization potential and ionization energy of the one-electron hydrogen atom. The value of λ ascertainable from its spectrum, is 912Å. This is equivalent to 9.12×10^{-6} cm, since $1\text{Å} = 1 \times 10^{8}$ cm. Putting all of our information together, we obtain for the hydrogen atom its ionization energy as follows:

$$E_{per\ atom} = h\nu = hc/\lambda$$

$$= \frac{6.63 \times 10^{-27}\ \text{erg·sec} \times 3.0 \times 10^{10}\ \text{cm/sec}}{9.12 \times 10^{-6}\ \text{cm}}$$

$$= 2.18 \times 10^{-11}\ \text{erg/atm.}$$

Consequently,

$$\text{ionization energy} = (2.18 \times 10^{-11}\ \text{erg/atm}) \times (1\ \text{kcal}/4.19 \times 10^{10}\ \text{ergs})$$

$$= 5.21 \times 10^{-22}\ \text{kcal/atm.}$$

This corresponds to

$$(5.21 \times 10^{-22}\ \text{kcal/atm}) \times (6.023 \times 10^{23}\ \text{atm/mole})$$

which yields 313 kcal/mole.
Likewise, we would have derived in electron volts (eV)

ionization potential = $(2.18 \times 10^{-11}$ erg/atm$) \times (1$ eV$/1.60 \times 10^{-12}$ erg$)$

= 13.6 eV/atm.

The very considerable quantity of electrical energy required to remove the electron from each atom of an entire mole of hydrogen atoms would be

$(13.6$ eV/atm$) \times (6.023 \times 10^{23}$ atm/mole$)$

which yields 8.19×10^{24} eV/mole.

ELECTRON AFFINITY

The electron affinity of an atom represents its energy as a neutral *isolated* gaseous entity to pick up an additional electron and form a monovalent negatively-charged ion. As in so doing a more stable entity is formed, the amount of energy that is necessarily released becomes an indicator of the tightness with which the atom of interest binds to itself the added electron. Unlike the ease with which ionization potentials are experimentally measured, the difficulties besetting any direct determination of electron affinity have permitted such assignments of energy to extremely few atoms. Some indirect theoretical extrapolations can be quantitatively made when atomic size and effective nuclear charge are carefully considered, but the values lack the authority of experimental reliability.

Electron affinity is sometimes confused with "electronegativity." Although a valid and understandable relationship exists between the two concepts, each offers a distinctly different theoretical approach to the evaluation of electronic attractions.

ELECTRONEGATIVITY

The concept of *electronegativity* evaluates the attractions being competitively exerted by two net-neutral atoms in a stable molecule for the *greater* share of a pair of electrons available to both and constituting their mutual chemical bonding. Note carefully the distinction from *electron affinity*, which concerns itself solely with the isolated atom and its complete and total acquisition of an electron which it draws to itself from a presumed infinite distance away. The attraction in electronegativity, on the other hand, is the intrinsic energy not of completely independent atoms but rather of the changes in such intrinsic energy wrought by incorporating atoms within stable molecular associations wherein they are now mutually interdependent.

ATOMIC PROPERTIES

In the sequence of their atomic numbers, the electronegativity of elements follows the same pattern of periodicity as do other properties associated with the general arrangements of the periodic table. It is thus possible to predict therefrom the relative reluctance with which atoms of the elements give up their valence electrons. Inspection of the Periodic Table (Appendix D) for the values for electronegativity reveals that in any periodic row of elements the figures for electronegativity progressively increase from left to right, and progressively diminish when going downwards in any particular group. These parallel, respectively, the diminution in the reducing capacity of the elements (ability to *lose* electrons to form positively-charged ions) in proceeding from left to right along a periodic series, and the increases in their reducing capacity in proceeding downwards in a group. As metallic characteristics also follow a trend in general accord with the periodicity of electronegativity, the most metallic of the elements are at the left of any series, and at the bottom of any group. That is, the associated general capacity in families of elements to form *acidic* compounds diminishes for the individuals thereof in going from the top downwards in their particular family group.

ASSIGNMENTS OF VALUES FOR ELECTRONEGATIVITY

The electronegativity scales that have been constructed from the various types of experimental data are completely arbitrary. As the origins are arbitrary, the units likewise are arbitrary.

Pauling Scale

The values of the Pauling scale are based upon the concept of bond energies. *Bond energy* may be defined simply as the change in energy involved in the binding of two atoms to form a stable molecule. As conceived when two different atoms A and B bond to each other, the change in energy involved in the formation of this covalent bond exceeds the geometric mean of the respective energies of the bonds A—A and B—B. This excess quantity of energy is determined to be proportionate to the difference between the respective electronegativity of the separate atoms, A and B.

Symbolizing electronegativity by X and bond energy by E, we obtain the increment of energy ΔE the following relationship:

$$\Delta E = X_A - X_B = E_{A-B} - \sqrt{E_{A-A} \times E_{B-B}}.$$

This formulation is a modification of the evaluation

$$(X_A - X_B)^2 = E_{A-B} - \tfrac{1}{2}(E_{A-A} + E_{B-B})$$

or

$$X_A - X_B = \sqrt{E_{A-B} - \tfrac{1}{2}E_{A-A} - \tfrac{1}{2}E_{B-B}}.$$

Mulliken Scale

Another approach to assignments of electronegativity is that of the Mullikan scale. On this scale the electronegativity of an atom reasonably approximates that of the Pauling scale when the *difference* (in *electron volts*) between the particular atom's ionization potential (*ip*) and its electron affinity (*ea*) is divided by a scale-adjustment factor of 5.6. Thus, for any atom, the measure in electronegativity, of its attraction for electrons is given by

$$X = (E_{ip} - E_{ea})/5.6.$$

Let us reflect a bit longer on the implications of bonding between the atoms A and B in forming a stable molecule by the sharing (covalence) of an electron-pair. There are three different possibilities, depending upon the relative competitive pulling power of each of the atoms for the shared electrons. Thus, with δ representing the *increment* of geographically displaced electrical charge in the net-neutral molecule AB, the following apply:

(1)	(2)	(3)
δ⁻ δ⁺	δ° δ°	δ⁺ δ⁻
A$_\times^\times$ B	A $\times\atop\times$ B	A $\times\atop\times$B
(1) A⁻—B⁺	(2) A°—B°	(3) A⁺—B⁻.

Figures 1.1 and 1.2 indicate the different polarities of the bonds that result from shifts of the electron-pair (indicated by crosses) in response to the alternate disproportionate electron attractions of the atoms A and B for electrons. Figure 1.2 represents a bond that is 100% covalent; that is, the sharing of the electron-pair between A and B is precisely equal. This condition of a perfect covalent bond, as contrasted with the polar bonds of Fig. 1.1 and Fig. 1.3, would require that any difference in the *absorption* of energy expressed by the ionization potentials of atoms A and B be compensated by an equal and opposing difference in energy *release* as expressed by the electron affinities of B and A.

By established custom, negative values of energy signify liberation and positive values signify absorption of energy. With electron affinities denoted as algebraically-negative energy values, therefore, we may equate the pertinent relationships as follows:

$$[E_{ipA} - E_{ipB}] = [-E_{eaB} - (-E_{eaA})],$$

which rearranges to

$$[E_{ipA} - E_{eaA}] = [E_{ipB} - E_{eaB}]$$

whereupon for a *purely* covalent bond without the slightest polarity,

$$[E_{ipA} - E_{eaA}] - [E_{ipB} - E_{eaB}] = 0.$$

The equality of electronegativity of A and B (i.e., $X_A = X_B$) is expressed on the Mulliken scale as

$$\left[\frac{E_{ipA} - E_{eaA}}{5.6}\right] - \left[\frac{E_{ipB} - E_{eaB}}{5.6}\right] = 0.$$

The condition of complete nonpolarity of bonding always prevails in a *homonuclear* (same chemical nucleus) diatomic molecule (e.g., H_2, Cl_2) inasmuch as a difference in either ionization potential or electron affinity can hardly be ascribed to one or the other of the identical atoms in the respective molecule. Where, however, *heteronuclear* molecules (different nuclei) are involved, as with A and B, if the numerical value of $(E_{ipA} - E_{eaA})/5.6$ be greater than that of $(E_{ipB} - E_{eaB})/5.6$ the bond created is that of the *polar* representation A⁻—B⁺. On the other hand, if the numerical value of the quotient pertinent to A be less than that of B (i.e., $X_A < X_B$) the polar bond created is that of A⁺—B⁻. It is consequently manifest that any differences respecting the polarity or nonpolarity of bonds — or what amounts to the same thing, the percentage of ionic character of covalent bonds — are all strictly matters of degree and not of intrinsic nature.

ORIGINS OF ATOMIC SPECTRA — DETERMINATIONS OF ATOMIC NUMBERS

Elements may be identified, and their atomic numbers established, by x-ray spectroscopy. Spectral lines result when a target anode made of the element of interest is subjected, in a vacuum tube, to bombardment by a stream of high-velocity electrons being emitted from a heated cathode within the tube. These highly energetic electrons transfer their energy upon impact and as a consequence, eject electrons from the inner electron shells of the target atoms. Upon removal of an electron from the K shell of an atom, the effective nuclear charge of the atom increases and pulls an electron in the L shell into the newly vacated position within the lower-energy K level. The initial extra energy of the L electron is thereupon released in the x-radiation characteristic of the K line in the spectrum of the target element. One may now visualize a sequential series of such x-radiations of progressively increasing wavelengths, wherein an electron from the M level drops into the vacant spot of the L shell; and, in turn, an N electron drops into the M shell, etc. Presumably, an N electron may drop directly down into the K shell. Hence, the emission spectrum of an element is describable in terms of a series of lines — K, M, N, O, ... etc.

The designations α, β, γ, and δ are normally appended as subscripts in order to define the precise type of transition. Thus, K_α, K_β, K_γ, K_δ, respectively, denote spectral lines associated with transitions to the K shell of electrons from the L, M, N, and O shells. Similarly, L_α, L_β, and L_γ express transitions to the L shell of electrons from the M, N, and O levels, respectively; and M_α, M_β transitions to the M shell of N and O electrons, respectively. See Figure 1.2.

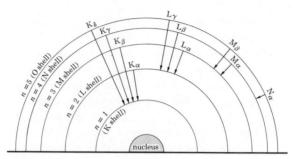

FIG. 1.2 Shell arrangement of an atom.

It must not be supposed, however, that spectral radiations are limited to the displacements of electrons solely between these principal energy levels. Literally many hundreds of lines in a spectrum are ascribable to the shifting of electrons from one sublevel of energy to another in the same principal energy level, each line always corresponding to a specific wavelength of light. When an atom is in an excited state, its electrons absorb energy and undergo displacements toward energy levels farther away from the nucleus. In all cases, whether radiated or absorbed, the amounts of energy involved are specific *quanta* defined by

$$(E_{\text{original}} - E_{\text{final}}) = h\nu.$$

The terms h and ν have already been explained. Clearly, when E_{original} is greater than E_{final}, energy is radiated; when E_{original} is less than E_{final}, energy is absorbed.

A quantitative relationship has been established between the frequency, ν, of a particular line in the spectrum of an element and its nuclear charge, Z (atomic number); namely, that the square root of the frequency is a linear function of the atomic number. This is known as Moseley's law, expressed mathematically as

$$\sqrt{\nu} = a(Z - b),$$

wherein ν is itself proportional to the reciprocal of the wavelength λ of the spectral radiation. The terms a and b are parameters whose numerical values are ascertainable by observations upon elements of known nuclear charge. The atomic numbers of all elements can thus be determined by experiment; and consequently, new elements are identifiable.

EXTRANUCLEAR NATURE OF THE ATOM; QUANTUM MECHANICS OF THE ENERGIES AND DISTRIBUTIONS OF ELECTRONS

Our concern now is with the energy that the atom contributes to chemical reactivity, expressed as an attribute of the number and the collective configuration of its electrons. (These interpretations are placed to particular use in Chapter 3, dealing with the bondings and structures of the transition complexes.)

The energy of the electron is *quantized* — that is, its specific discrete value can be assessed in terms of its orientation with respect to the nucleus of the atom and to the magnetic field to which it, as a spinning electrical charge, must contribute.

In conformity with the discontinuities in states of energy, excitation of an electron from a lower to a higher state of energy occurs only with an abrupt absorption of a discrete amount of energy that is precisely equal to the difference in the amounts of energy of the two states. Similarly, an electron that drops from a higher ("excited") to a lower state of energy must emit energy, likewise abruptly, in a discrete amount identical to that of the difference between the two states involved. It is this emission of energy by the electron as it returns to a more stable condition that accounts for *line spectra*, because the energy is radiated in the form of light of characteristic frequency.

If the electron is already in the ground state — that is, its minimum or lowest state of energy — it cannot emit energy. This is an answer to the question why electrons do not plummet into the nucleus, causing collapse and annihilation of the atom. It, likewise, justifies the rejection of the classical but highly fallacious comparison of the motion of an electron with that of a planetary body revolving in a fixed orbit around the sun. Any electrically charged particle moving in a fixed path under the influence of attractional forces such as exists between the atom's negatively charged electrons and the positively charged nucleons — must radiate energy. Hence, such a construction of the atom inevitably contradicts the very existence of matter.

It must be noted, however, that the foregoing considerations do not preclude the capture by an *unstable* nucleus of an electron from the K-shell, or even from the L-shell, of an atom ("K-capture," "L-capture"). This occurs in radioactive transitions of unstable nuclei whose neutron/proton ratios lie below the belt of stability. If this belt of stability be regarded as a numerical equality of neutrons to protons — that is, of neutrons/protons $= 1$ — then, for the condition neutrons/protons < 1, an atom increases its stability by taking a near electron and converting it to a neutron by interaction with a proton. This may be symbolized in equation form, as follows,

$$_{+1}p^1 + {}_{-1}e^0 \rightarrow {}_0n^1$$

in which superscripts denote masses of particles, and subscripts the charges on particles.

Upon occurrence of this change, x radiation is emitted as external electrons drop into vacated places within the lower orbital energy levels. In any event, whether existing with stability or in the process of radioactive transformation, the electrons of atoms remain *quantized*.

The original Bohr theory (Niels Bohr, Denmark, 1913), of the electron as a corpuscular particle, to which modern-day quantum mechanics defers in its generalizations regarding the quantized restricted total energies of electrons, has become somewhat inadequate as an interpretative tool. Particularly unrewarding is any attempt to plot the path or trajectory of an electron in a given energy level. Particles of very small mass and charge do not respond in the same manner to classic laws of gravitational motion (Newton's) and electrical interactions as do larger bodies. The *Heisenberg Principle of Uncertainty* informs us that it is apparently impossible to determine with accuracy the *energy content* (or *momentum*) of the electron and its *position* in the atom simultaneously. All experimentation that has so far been devised and tried to measure one, automatically induces an alteration in the other. Electromagnetic radiation, which might otherwise provide the image of the electron, cannot be applied reliably inasmuch as the dimensions of an electron are smaller than the wavelength of such radiation. Likewise unreliable proves the application of any radiation of wavelength less than the dimensions of the electron; the greater energy of such radiation then profoundly alters the energy of the electron. In the relationship,

$$E = \frac{hc}{\lambda} = h\nu$$

an acceptably reliable numerical value for the proportionality constant h, (Planck's constant, in erg·sec) would appear to be the uncertainty factor.

Thus it is to be recognized that the properties of electrons are best described and evaluated in terms of their wave motions. Hence, we conceive of the electron, instead, as a cloud or region, within the atom or molecule, of variable charge density corresponding to its assessed *probability* of being found or located.

CONTRIBUTIONS OF DE BROGLIE AND SCHRÖDINGER TO QUANTUM MECHANICS

The wave-concept of material particles of very small mass was first suggested by Louis de Broglie (France, 1924). The subsequent statistical mathematics of a *wave equation*, introduced by E. Schrödinger (Austria, 1926) permits the probabilities of location and properties of each electron in an atom to be defined as a wave function condition of four quantum numbers.

Certain qualitative interpretations of the significance of the Schrödinger equation are in order inasmuch as this equation represents both the foundation for the larger portion of our concepts of quantum mechanics and the computational basis for deriving the four quantum numbers. The analogy of the properties of the electrons to the properties of waves postulates only a statistical approach to resolving the difficulties in ascertaining concomitantly both the position and the momentum of an electron, yet it remains our present sole recourse to acceptable quantitative evaluations.

Let us see how the Schrödinger equation looks. Based fundamentally upon the de Broglie postulate that every moving particle has the characteristics of waves — that is, it may be described in terms of the wave frequency established by the velocity of the particle — one of the several arrangements of the equation takes the following form:

$$\left(\frac{\partial^2 \psi}{\partial x^2} + \frac{\partial^2 \psi}{\partial y^2} + \frac{\partial^2 \psi}{\partial z^2}\right) + \frac{8\pi^2 m}{h^2}(E - U)\psi = 0.$$

In this, x, y, and z, are the familiar Cartesian coordinates and

$m =$ mass of moving particle,

$\pi = 3.1416$,

$E =$ total energy of particle — potential energy, U, plus kinetic energy as defined in $\frac{1}{2}mv^2$,

$h =$ Planck's constant,

$\psi =$ wave function of moving particle; equivalent to amplitude of de Broglie wave.

This value of h represents, in itself, the *minimum* uncertainty (as described by the Heisenberg Principle of Uncertainty) of defining simultaneously the electron's momentum and its location, expressed in the form

$$(\Delta p)(\Delta x) \geq h$$

wherein Δp denotes the uncertainty of the momentum and Δx the uncertainty of the location with respect to a distance x.

The importance of ψ in the equation is not in itself, because it lacks any true physical significance; rather with its square; for it is acutally ψ^2 that expresses the distribution of the probability cloud or relative density of charge of the electron wave. The larger the value of ψ^2, the greater the chance of finding the electron.

With the exception of E and ψ, all factors in the equation are known. The solution of this otherwise formidable equation for ψ responds only to certain allowable values of E — all of which are *integrally* related.

ATOMIC PROPERTIES

It is possible to differentiate the over-all three-dimensional Schrödinger equation into the three separate differential components representing the motion of the probability cloud along each of the separate Cartesian coordinates. Thus, illustratively, the motion of the electron in the direction of the y axis would respond to the one-dimensional formulation:

$$-\frac{h^2}{8\pi^2 m}\frac{d^2\psi}{dy^2} + U\psi = E\psi.$$

THE FOUR QUANTUM NUMBERS

1. n, *principal* quantum number.

The values assigned to n designate the average distance of the electron from the nucleus of the atom as it (the electron) contributes to the probable spatial distribution of the over-all electron cloud comprising a particular main level of energy. Thus, $n = 1$, $n = 2$, $n = 3$, $n = 4$, etc., represent progressively increasing main energy levels of spatial separation of the electron from the nucleus.

2. l, *orbital* (azimuthal) quantum number.

The values assigned to l designate the radial or angular momentum of the electron and define the particular shape of the atomic orbital. An *orbital* is to be regarded as a three-dimensional locus or orientation in space where the probability is greatest for finding *one* electron *alone*, or *two* electrons *paired*, as a consequence of the electromagnetic attraction they exert for each other when their spins are opposed, clockwise and counterclockwise. The shapes that are given to orbitals represent surface-connected boundary points of three-dimensional areas of greatest probability of electron density — roughly 90% or so of the effective charge cloud (Fig. 1.3). Differently phrased, the shapes of orbitals outline approximate regions where, in accordance with probability factors, the electron may be presumed to spend about 90% of its time (See next section).

3. m_l, *magnetic* quantum number.

The values assigned to this quantum number identify the orientations of an electron cloud in a magnetic field. Quantitatively, the quantum number herein accounts for the splitting of spectral lines (Zeeman effect) when emitted electrons — themselves, tiny magnets — traverse an externally applied magnetic field. The following magnetic quantum values are assigned to the magnetic-field orientations of the orbitals:

$m_l = 0$; s orbital, only 1 value possible,

$m_l = 0, +1, -1$; p orbital, any of 3 values,

$m_l = 0, +1, +2, -2, -1$; d orbital, any 1 of 5 values,

$m_l = 0, +1, +2, +3, -3, -2, -1$; f orbital, any 1 of 7 values.

4. m_s, *spin* quantum number.

The values assigned to this quantum number indicate the direction of spin that orients the electron by virtue of the latter's characteristics as essentially a submicroscopic magnet. This quantum value for an electron can be either one of two solely regardless of orbital configuration:

$$m_s = +1/2, -1/2.$$

Two electrons having parallel spins must have identical values of m_s (either $+1/2$ or $-1/2$). Clearly, as similarly-oriented magnets, they cannot pair up. Should they have opposite spins — one with $m_s = +1/2$, and the other with $m_s = -1/2$ — they can form electron-pairs.

The four quantum numbers that have just been described justify the electron configurations assigned to each of the individual elements provided, however, that they are appraised in terms of certain additional restrictions of quantum mechanics being placed upon their exact numerical values. As defined, these restrictions are:

Pauli Exclusion Principle (W. Pauli, 1925). No two electrons in the same atom can have the same exact set of the four quantum numbers. This, quite logically, labels each electron in an atom with its own individual identification.

Hund's Rule of Maximum Multiplicity (F. Hund, circa 1928-29). Electrons in the same orbital state and unperturbed by the approach of ligands remain unpaired as long as possible. This is equivalent to a requirement that they then have parallel spins and normally enter each orbital of a specific subshell singly before any pairing can possibly occur between those of opposite spin. Hence, each orbital of any specific subshell holds one electron before any other orbital in that specific sublevel may acquire a second. This conforms to experimental measurements (with *Gouy balance*) of the degree of *paramagnetism* of a species; that is, the weak magnetic attraction associated with the number of unpaired electrons held by a species, and which must overcome the *diamagnetism* or repelling effects of the electrons themselves. Paramagnetism results from the spin of unpaired electrons; diamagnetism from their similarity of *electrical charge*. When electrons are paired, each cancels out the effective magnetic properties of the other; hence, the contribution to paramagnetism of an electron-pair is zero.

SHAPES OF ORBITALS

It must be borne in mind that the boundary surfaces of orbitals connect points of greatest electron density. The one electron or the two electrons (as pertinent) being viewed as an electron cloud, must be presumed to occupy each and all of the separate lobes of the orbital simultaneously; this, by virtue of the electron's kinetic, translational energy.

The following rules define the values of the orbital quantum number, l, described in the preceding section:

1. If the individual atomic orbital is *spherical*, $l = 0$.

These orbitals are described as s orbitals, the electron(s) therein as s electrons, and the collective areas of their density within the main energy levels as s sublevels, subshells, or subgroups. [See Fig. 1.3, (a).] The designation s actually refers to the spectroscopist's identification of the lines in the spectrum as *sharp*. For any specific main energy level there can be only one s orbital or subshell and, diagrammatically, portrayed as a sphere centered at the origin of three mutually perpendicular Cartesian axes.

2. If the individual atomic orbital is *two-lobed* and like a dumbbell, $l = 1$.

These are the p orbitals and their electrons and energy-level subgroups; the p refers to the designation of spectral lines as *principal*. For any given main energy level *beyond the first* there are three p orbitals available, accommodating a maximum total of six electrons (two in each p orbital). [See Fig. 1.3, (b).] There can be no p orbitals or electrons whatsoever in the first main energy level. All three p orbitals must have identical energies but separate identities as individuals. This precludes any representation of each as a sphere, inasmuch as all three p orbitals would then have in conformity with identical energies a common coordinate center and an identical diameter. That is, it would yield a net effect of just one orbital with six electrons. This, consequently, is totally precluded as a possibility because of the restriction placed upon the maximum number of electrons in any single orbital, of whatever desigation, to not more than two electrons. The diagrammatic representations of the three separate p orbitals, take courses along the three Cartesian axes, x, y, and z. However, the three individual orbitals collectively considered as superimposed may be totally visualized as concentrically spherical around the origin of the Cartesian axes.

3. If the individual atomic orbital is *four-lobed*, likened to two crossed dumbbells $l = 2$.

These identify d orbitals [Fig. 1.3 (c)]. A hybrid modification, to which the quantum number, $l = 2$, is also assigned, may best be described as one dumbbell with a doughnut-shaped ring concentrically arranged perpendicular to its center. These identify the d orbitals (as the spectral lines charac-

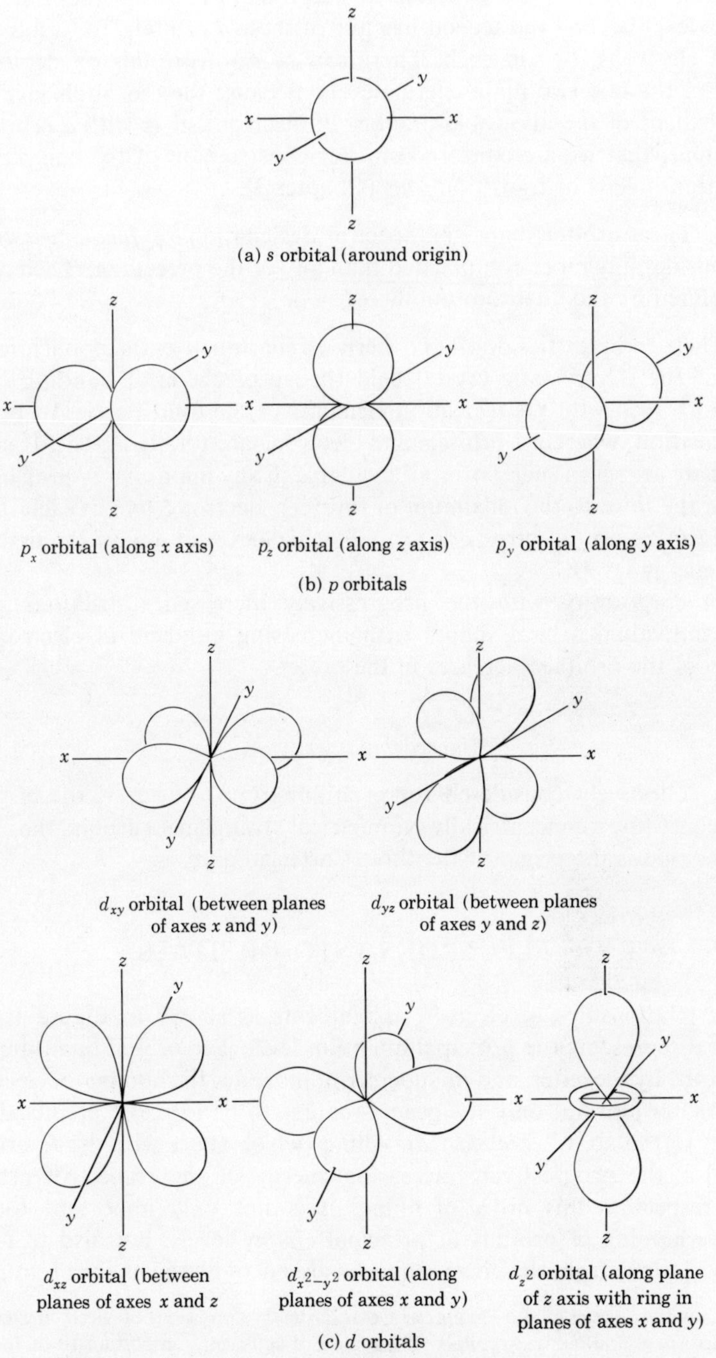

ATOMIC PROPERTIES 19

(a) *s* orbital (around origin)

p_x orbital (along *x* axis) p_z orbital (along *z* axis) p_y orbital (along *y* axis)

(b) *p* orbitals

d_{xy} orbital (between planes of axes *x* and *y*) d_{yz} orbital (between planes of axes *y* and *z*)

d_{xz} orbital (between planes of axes *x* and *z*) $d_{x^2-y^2}$ orbital (along planes of axes *x* and *y*) d_{z^2} orbital (along plane of *z* axis with ring in planes of axes *x* and *y*)

(c) *d* orbitals

FIG. 1.3   Orbital arrangements of atoms.

terized as *diffuse*) and the electrons of which they are comprised. Each main energy level beyond the second has five of these *d* orbitals, for a maximum of ten electrons, two in each. There can be no *d* orbitals or electrons in either of the first two main energy levels. Because these orbitals lack a full complement of the allowable number of electrons, it is with *d* orbitals, in particular, that we are concerned in our construction of the complexes of transition metals or transition ions (Chapter 3).

4. The *f* orbital shapes (*f*, spectral designation of *fundamental* lines) are considerably more complicated than any of the preceding. Their shapes are defined by the quantum number, $l = 3$.

Their complexities do not concern us, inasmuch as the applications we seek of the electrostatic crystal field theory of chemical bonding can be best portrayed with the transition elements of the fourth series of periodic classification, wherein *f* orbitals and electrons are totally absent. It suffices that there are seven such *f* orbitals available in any main energy level beyond that of the third to the maximum of fourteen electrons, two in each orbital. There can be no *f* electrons or orbitals whatsoever in any of the first three main energy levels.[1]

In conformity with the progressively increased restrictions placed upon individual orbital shapes with increasing numbers of electrons the energy of the orbitals increases in the order

$$s < p < d < f \quad \dots\dots\dots\dots\dots\dots$$
increasing energy →

When collectively considered for each subgroup, however, the electrons must constitute a concentrically symmetrical arrangement around the atomic nucleus that is the origin of the three Cartesian axes.

ENTRANCE OF ELECTRONS INTO ORBITALS

In the relationships of electron assignments as shown in Figure 1.4, the numeral represents the principal quantum level. The orbital quantum level is denoted by the letter, and its superscript indicates the number of electrons. The arrows plot not only the general course to be logically predicted with respect to the atom's *ground-state* filling (which starts with the *1s* orbital), but, also, the progressively increasing energy of the respective orbitals. With respect to this order of filling, it is not only important to note the overlapping of orbitals of principal energy levels, but also to realize that deviations from the given sequence are encountered as the atom grows

---

[1] Quantum values of $l > 3$ represent extraordinary conditions of electron excitation presumed for *g* and *h* ..... orbitals. These are not sufficiently important to be included in the scope of required structural development being considered here.

ATOMIC PROPERTIES

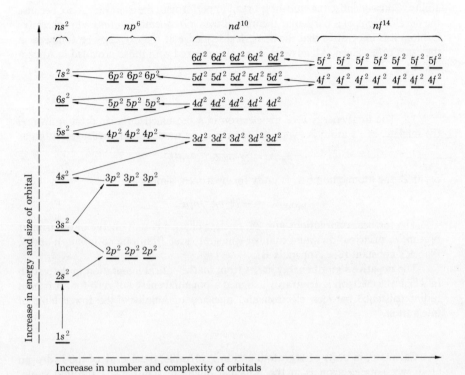

FIG. 1.4 Ground-state energy and maximum filling of orbitals.

in complexity. Thus, the 4f and 5d orbital energies are so very close to one another that inversions of the orbitals may result. Even with lower energy states, the regularity of ground-state filling of orbitals may be altered. Such adjustments are generally attributed to the greater stabilities of energy sublevels when their respective orbitals are *all* exactly half-filled. The order of the ground-state fill-in that has been charted in Figure 1.4, however, remains quite adequate to cope with ground-state requirements of all elements known.

EXERCISES

Note: Many of these exercises have been planned to provide important new factual information while at the same time seeking tests of understanding and memory of material already presented within the chapter context as well as of mathematical ability. Consequently, the student is urged to perform all the exercises. Also, because the development of a particular theoretical theme frequently continues in successive fashion, it is imperative that the numerical sequence of the exercises be followed in the exact order given. All answers should be checked with those provided in Appendix A.

1. The total energy E of the electron in a presumed circular orbit n around the nucleus of its atom is expressed for the Bohr model of the hydrogen atom as

$$E_n = -2\pi^2 m(Ze)^2 e^2/n^2 h^2$$

or, as Z, the atomic number, is unity for hydrogen, simply as

$$E_n = -2\pi^2 m e^4/n^2 h^2.$$

The further connotations are: Ze, nuclear charge; e, unit charge of electron or proton; n, principal quantum number (integral value from one to infinity); and h, Planck's constant (See Appendix B).

The negative sign of energy arises from mathematical integration in derivation in which the electron is arbitrarily assigned a potential energy of zero for an orbital radius (distance between electron and nucleus) of infinity — the lower limit of integration.

Calculate, in units of erg/atom and in kcal/mole, the energy of the hydrogen atom when its electron is in the principal quantum state, $n = 1$ (ground state).

2. The energy of a quantum of light ϵ emitted when an electron is displaced from a higher orbital level n_2 to a lower orbital level n_1 is the numerical difference between the respective energies of the two levels; namely

$$\epsilon\,(= h\nu) = E_{n_2} - E_{n_1}.$$

The energies of the pertinent levels are equated as

$$E_{n_1} = -2\pi^2 m(Ze)^2 e^2/n_1^2 h^2$$

and

$$E_{n_2} = -2\pi^2 m(Ze)^2 e^2/n^2 h^2.$$

Write a *combined* general expression that equates the energy of the emitted radiation ϵ to the principal quantum states, n_1 and n_2.

3. Fundamental relationships are

ATOMIC PROPERTIES

$$\epsilon = h\nu$$

and

$$c = \lambda\nu$$

wherein ϵ is the energy of a quantum of light radiation; λ and ν the wavelength, and the frequency of the light, respectively; c the velocity of light; and h the Planck constant. These relationships may be employed to equate the wave *number* $\bar{\nu}$ (number of waves to centimeter; hence, reciprocal of wavelength, $1/\lambda$) as follows:

$$\bar{\nu} = \frac{1}{\lambda} = RZ^2\left(\frac{1}{n_1^2} - \frac{1}{n_2^2}\right)$$

wherein R is the *Rydberg* constant of proportionality.

(a) Evaluate R in the general terms of m, e, h, and c.
(b) Utilizing the pertinent fundamental constants given in the Appendix, calculate to four significant figures the numerical value of R in units of (1) cm^{-1}; (2) Å^{-1}.

4. To ionize a hydrogen atom in accordance with the equation,

$$H^0_{(g)} \rightarrow H^+_{(g)} + 1e^-$$

the energy absorbed must be sufficiently large, that is, the wavelengths of radiation must be sufficiently short to raise the electron from its initial principal quantum level to an orbital level of $n = \infty$. Beyond this upper limit of energy defined by $n = \infty$, the hitherto discrete lines of the spectral radiation that characterize a particular series of lower quantum levels will have vanished and a *continuum* (a continuous spectrum devoid of discernible discrete line components) will have commenced, wherein radiations of all wavelengths in the region of the continuum are absorbed. In this region, the energy of the electron can no longer be regarded as quantized.

(a) Calculate the wavelength λ (in angstroms) of the series limit corresponding to radiation emitted by the hydrogen atom during the transition of electrons from $n = \infty$ to its ground state of $n = 1$, in accordance with the relationship

$$\frac{1}{\lambda} = RZ^2\left(\frac{1}{n_1^2} - \frac{1}{\infty}\right).$$

(b) Calculate the wavelength of radiation corresponding to transition of the electron from its continuum $n = \infty$ to the excited state $n = 2$.

24   ATOMIC PROPERTIES

5. Calculate
(a) the heat energy in erg/atm,
(b) the heat energy in kcal/mole,
(c) the electrical energy in electron volts/atom, that must be absorbed by the hydrogen atom whose electron is in the ground state, $n = 1$, in order to effect complete ionization. (Recall that ionization of the atom requires removal of the electron to $n = \infty$, where it is determined that $\lambda = 911.6$ Å).

The interrelationships $\dfrac{1}{\lambda} = \dfrac{\nu}{c} = \dfrac{E}{hc}$ likewise prove applicable to the solution of the problem.

6. The equation for determining the hydrogen atom's energy in any principal quantum number of its electron has already been provided (Exercise 1). It is apparent from inspection that once the energy has been established for the orbital ground state of $n = 1$, the energy of the atom in excited states, that is, principal quantum numbers of any integral values from 2 to ∞ — may be quickly ascertained by multiplying the energy value that corresponds to $E_{n=1}$ by the reciprocal of the square of the quantum number of the pertinent excited state, ($1/n^2$ excited state).

(a) Calculate to three significant figures the energy of the hydrogen atom, in ergs/atom and in kcal/mole, corresponding to the excited quantum states, $n = 2$; $n = 3$; $n = 4$; $n = 5$.

(b) Why are all values of E_n negative? What will be the maximum value of E_n in a quantized state of the atom?

7. Line emission spectra for the Bohr model of the hydrogen atom are represented by several series, each dependent upon the specific series *initial base* quantum number (n_1) whence all excitation of the electron to *higher final* energy levels (n_2) occurs. All series culminate in an identical upper limit of line spectra defined by $n = \infty$, which represents the inception of the spectral continuum of the atom's complete ionization. The applicable formula for the determination of λ (radiation wavelength) has been given as

$$\frac{1}{\lambda} = RZ^2\left(\frac{1}{n_1^2} - \frac{1}{n_2^2}\right)$$

with n_1 necessarily denoting the particular series base and $n_2 > n_1$.

The following data is pertinent to our immediate interest:

Hydrogen Spectral Series	Region of Line Spectra	Series Base
(e) Pfund	infrared	$n_1 = 5$
(d) Brackett	infrared	$n_1 = 4$
(c) Ritz-Paschen	infrared	$n_1 = 3$
(b) Balmer	visible	$n_1 = 2$
(a) Lyman	ultraviolet	$n_1 = 1$

Calculate for each of these series the wavelength (in angstrom units) of the very first spectral line in it which represents the energy radiation of the electron in dropping down one quantum unit to its respective series base. Take $R = 1.097 \times 10^5$ cm^{-1}.

8. In the equation employed in the preceding exercise why cannot the quantum value of n_2 be less than or equal to that of n_1?

9. (a) The spectra of He$^+$ ($Z = 2$), Li^{2+} ($Z = 3$), and Be^{3+} ($Z = 4$) are very much like that of H^0 ($Z = 1$). Why?
(b) How will the frequencies ν and the wave numbers $\bar{\nu}$ (the number of waves to cm) of He$^+$, Li^{2+}, and Be^{3+} compare, approximately, with those of H^0?
(c) How will the radiation wavelengths λ of He$^+$, Li^{2+}, and Be^{3+} compare, approximately, with that of H^0?
(d) Calculate λ, in angstroms, of the first spectral lines in the Balmer series for He$^+$, for Li^{2+}, and Be^{3+}, respectively.

10. What explanation interprets the facts that the Cr^0 atom is defined electronically by the orbital designation [Ar]$3d^5\ 4s^1$, rather than by [Ar] $3d^4\ 4s^2$; and that Cu^0 is likewise defined by [Ar] $3d^{10}\ 4s^1$, rather than by [Ar] $3d^9\ 4s^2$?

11. Why will electrons enter singly into each orbital of a given energy sublevel? And why is electron-pairing at all possible?

12. The following interpretation may be made of the calculated approximations that constitute the probability distributions of electrons in their respective orbitals and consequently, of the shapes of such orbitals. Despite the progressively increasing elongations of individual p, d, and f orbitals with respect to their orientations about the atom's nucleus (s orbitals are always individually spherical, unless distorted by other nearby orbitals), an empirical rule that may be applied is that when each orbital of any given subshell is occupied by an identical number of electrons (two or one, or zero in each) their over-all net summation yields a subshell that is symmetrically spherical around the nucleus.

For each of the following series of species variations give the orbital distributions of the pertinent electrons that substantiate a stated correct conclusion as to which of the individuals therein yields an over-all spherical and which a nonspherical electron cloud. Refer to the Periodic Table (Appendix D) for applicable data.

(a) Cu^0, Cu^+, Cu^{2+}
(b) N^0, N^{3-}
(c) Mn^0, Mn^{2+}, Mn^{3+}
(d) Fe^{2+}, Fe^{3+}
(e) Ga^0, Ga^{3+}
(f) Ne^0, Ne^+, Ne^{3+}.

13. The effect of nuclear charge upon the strength of the binding of an electron to the nucleus of its atom may logically be appraised for identical values of the principal quantum number n by considering isoelectronic species; that is, atoms or simple ions having identical electronic structures (numbers and orbital configurations of their electrons being precisely the same). Under such conditions, the otherwise variable factor of electron–electron repulsion is equalized for all and maintained constant for comparative purposes, thus allowing the factor of nucleus-electron attraction to be qualitatively compared without undue difficulties. As the energy of the one-electron hydrogen atom measures, in total effect, the binding energy of the atom's single electron to its nucleus of single protonic charge, it follows that for all one-electron species that differ only in nuclear charges (atomic number Z) the strength of nuclear binding of the electron may be calculated as the energy that must be absorbed to break the bond. For a one-electron species this has already been determined by the Bohr formulation provided for the hydrogen atom; namely

$$E = -2\pi^2 m Z^2 e^4 / n^2 h^2.$$

Given the ionization potential of gaseous H^0 ($H^0 \rightarrow H^+ + 1e^-$) as 13.6 eV, calculate from this the respective ionization potentials of the following gaseous species, all of which are isoelectronic with the hydrogen atom ($1s^1$):

He^+, Li^{2+}, Be^{3+}, B^{4+}, C^{5+}.

14. With respect to the preceding exercise: (a) write a chemical equation for each of the individual ionizations and designate the order of magnitude (first, second, third, etc.) of ionization potential of the gaseous neutral atoms, He^0, Li^0, Be^0, B^0, and C^0 that individually correspond to the respective changes described by the equations. (b) calculate the energy of the electron in the boron ion B^{4+} in kcal/mole.

15. The electron affinity of the nitrogen atom is evaluated as an energy release of equivalent to 0.0 electron volts. The carbon atom immediately preceding it in the periodic series, however, and the oxygen atom immediately following it, are both assigned rather sizeable electron affinities. Explain this anomaly.

ATOMIC PROPERTIES

16. The diameter of the nucleus of a given atom is 3.20×10^{-13} cm; the total volume of the entire atom is 1.14×10^{-22} cc. Assuming completely spherical forms both for the atom as a whole and for its nucleus, calculate with respect to it,
    (a) nuclear volume,
    (b) atomic diameter,
    (c) numerical ratio of nuclear radius to atomic radius,
    (d) percentage of space (volume) occupied by the nucleus within its atom.

17. Given the following relationships:
velocity = distance/time = cm/sec
energy = mass × (velocity)² = grams × (cm/sec)², and
the de Broglie definition of a particle as $\lambda mv = h$
wherein λ is wavelength, m is mass,
v is velocity, and h is Planck's constant.
Calculate
    (a) the de Broglie wavelength associated with an electron traveling at 4.60×10^8 cm/sec. Use the rest mass of the electron (see Appendix B).
    (b) the de Broglie wavelength associated with a stationary electron.

18. Calculate the energy equivalent of a single atom of helium of mass 4.003 atomic mass units (amu)
    (a) in calories.
    (b) in MeV (million electron volts).

19. The existence of a number of the following orbitals must be excluded because they cannot possibly be reconciled with concepts of quantum mechanics. State the identity of each and your interpretations of the reasons for excluding them.

$2d, 3f, 4f, 4g, 5g, 5h, 6g, 6h.$

20. State the maximum possible number of electrons available in the total orbitals of
    (a) the principal quantum level, $n = 6$.
    (b) the $7g$ sublevel, if the incrementally acquired number of orbitals there responds to the identical arithmetical progression of orbitals already delineated for $s \ldots p \ldots d \ldots f$.

CHAPTER

TWO

PROPERTIES OF CHEMICAL BOND

FIRST, WE INVESTIGATE PROCEDURES FOR ASCERTAINING THE PARTIAL IONIC character of the polar covalent bond that results from the unequal sharing of a pair of electrons between two bonded atoms in a stable molecule.

BOND POLARITY AND IONIC CHARACTER

An empirical relationship has been approximated that integrates into a common value the different values for electronegativity (X) as found on the Pauling and Mulliken scales:

$$X \approx \frac{E_{ie} + E_{ea}}{30},$$

wherein the terms E_{ie} and E_{ea} are, respectively, the ionization energy and electron affinity of the given atom, with units expressed chemically in *kilocalories/mole*.

29

PROPERTIES OF CHEMICAL BOND

When applied to the two atoms in a polar covalent bond, the preceding relationship yields the following further approximation:

$$\% \text{ of ionic character} = 16(X_> - X_<) + 3.5(X_> - X_<)^2$$

wherein $X_>$ represents the larger and $X_<$ the smaller of the electronegativity values of the two respective atoms of the chemical bond.

This equation readily supplies the mathematical information that a bond that is 50% ionic (and also 50% covalent) corresponds to a numerical difference of approximately 2.1 units in the electronegativities of two bonded atoms in a stable molecule. Thus,

$$\approx 50\% \text{ ionic character} = 16(2.1) + 3.5(2.1)^2.$$

The satisfactions of confirmation are frequently the reassurances sought by serious students of their acquired understanding. We might therefore advantageously use the preceding derivations in two ways:

1. to calculate and compare with the recorded values for electronegativity assigned to the halogen atoms (see Periodic Table, Appendix D). In this manner we establish and confirm the sequence in which these elements progressively change in their over-all capacities to attract electron-pairs in their respective compounds.

2. to calculate the percentage of ionic character of each of the hydrogen halides and so confirm the sequence in which the polarities of the hydrogen-halogen bonds progressively change in response to changes in electronegativity.

Using units of *electron volts* and utilizing the recorded values (periodic table) of the first ionization potential and of electron affinity — algebraically denoted, as already stressed in accordance with established convention, by a positive value for the former (since energy is being absorbed) and by a negative value for the latter (since energy is being released) — we derive via the Mulliken relationship,

$$X = \frac{E_{ip} - E_{ea}}{5.6}.$$

The figures for electronegativity of all the atoms of interest herein are as follows:

$$X_\text{H} = \frac{13.6 - (-0.7)}{5.6} = 2.5 \text{ (compares with 2.1, Pauling)}$$

$$X_\text{F} = \frac{17.4 - (-3.7)}{5.6} = 3.8 \text{ (compares with 4.0, Pauling)}$$

$$X_\text{Cl} = \frac{13.0 - (-4.0)}{5.6} = 3.0 \text{ (compares with 3.0, Pauling)}$$

$$X_{Br} = \frac{11.8 - (-3.8)}{5.6} = 2.8 \; (\textit{compares with 2.8, Pauling})$$

$$X_I = \frac{10.4 - (-3.4)}{5.6} = 2.5 \; (\textit{compares with 2.5, Pauling})$$

We have thus established and confirmed that electronegativity of the halogen atoms progressively diminishes in the sequential order,

$$F > Cl > Br > I.$$

To determine the percentage of ionic character of the hydrogen-halogen bond in each halide, we employ the empiricized Pauling–Mulliken approximation,

% of ionic character $= 16(X_> - X_<) + 3.5(X_> - X_<)^2$

whereupon we obtain

% Ionic Character

HF
$$\begin{cases} = 16(X_F - X_H) + 3.5(X_F - X_H)^2 \\ = 16(4.0 - 2.1) + 3.5(4.0 - 2.1)^2 \\ = 43\% \text{ ionic } (\approx 57\% \text{ covalent}) \end{cases}$$

HCl
$$\begin{cases} = 16(X_{Cl} - X_H) + 3.5(X_{Cl} - X_H)^2 \\ = 16(3.0 - 2.1) + 3.5(3.0 - 2.1)^2 \\ = 17\% \text{ ionic } (\approx 83\% \text{ covalent}) \end{cases}$$

HBr
$$\begin{cases} = 16(X_{Br} - X_H) + 3.5(X_{Br} - X_H)^2 \\ = 16(2.8 - 2.1) + 3.5(2.8 - 2.1)^2 \\ = 13\% \text{ ionic } (\approx 87\% \text{ covalent}) \end{cases}$$

HI
$$\begin{cases} = 16(X_I - X_H) + 3.5(X_I - X_H)^2 \\ = 16(2.5 - 2.1) + 3.5(2.5 - 2.1)^2 \\ = 7\% \text{ ionic } (\approx 93\% \text{ covalent}). \end{cases}$$

Consequently, we establish and confirm that the percentage of ionic character of the hydrogen halide bonds progressively diminishes in the sequential order

$$HF > HCl > HBr > HI.$$

It is clear that the actual mathematical results obtained with the Pauling-Mulliken relationship must reflect the sometimes profound uncertainties involved in reconciling the different electronegativity values shown on the Pauling and Mulliken scales. Such reconciliations are necessarily crude because of the severe difficulties encountered in ascertaining the component electron affinity values. In fact, electron affinities have been worked out in acceptable fashion for extremely few of the elements. Add to this the further dilemma of deciding, for simplified mathematical work, which of the electronegativity scales to utilize for computative substitutions. With the halogens, at any rate, the situation is not critical inasmuch as identical sequences of bond polarities are obtained with either of the electronegativity scales despite the numerical differences (not serious here) between the respective computed electronegativity values. In making the foregoing calculations, we elected to substitute the Pauling figures because, for the halogen atoms at least, but by no means necessarily for other atoms, somewhat greater credence is lent to support of measured dipole moments and strength of bonds. These bond properties are discussed later.

It is clear, in any event, that the larger the value of $X_<$ assigned to an atom, reflecting increased ionization potential or electron affinity, the greater is the degree of covalence of its bond with a given halogen. Correspondingly, the smaller the value of $X_<$ assigned — reflecting decreased ionization potential or electron affinity — the greater is the percentage of its ionic character. We would expect, then, for gaseous molecules wherein atoms of high electron affinity are associated with atoms of low ionization potential (strongly electropositive) that the bonds would be on the verge of complete ionic character, regardless of the specific chemical identities of the pertinent atoms involved.

Thus, the representative elements of Groups I and II of periodic classification, because of their rather small ionization potentials, all strongly favor a virtually complete ionic condition of their chemical bonds. This characterizes their chemical disposition to form true salts as contrasted with those of other electropositive elements which very largely yield so-called pseudo salts of highly significant, sometimes extremely extensive, degrees of covalence.

Inasmuch as the preceding expositions have dealt with the gaseous molecule, we should now consider the validity of reference to a *molecular* bond between an atom that is very strongly electropositive (very low ionization potential) and an atom that is very strongly electronegative (very high electron affinity). It would be quite incorrect to speak of the existence of molecules of a true salt in its crystalline solid, fused (molten), and dissolved states. The x-ray pattern of crystalline NaCl, for example, reveals a stable three-dimensional structure wherein each Na^+ ion is bonded to six Cl^- ions and each Cl^- ion, in turn, is bonded to six Na^+ ions. Nonetheless, the examination of spectra of gaseous (vaporized) NaCl readily reveals evidence of the existence of discrete $[Na^+ Cl^-]$ units. Such discrete and independent units of consistent chemical identity validly conform both to our concept and to our definition of "molecule."

BOND STRENGTH

The strength of a bond is to be regarded as definitively identical to its energy. Bond energy is the energy required to rupture the covalent bond that holds together the atoms of an otherwise stable molecule, and thus to form the neutral atoms. Manifestly, such expenditure of energy must, in parallel, measure the mutual binding strengths attributable to the atoms of the given bond.

The strength of a covalent bond becomes greater as the extent of its polarity increases. As we have observed in the previous chapter, any shift of an electron-pair from the geographical center of separation of the two atomic nuclei is, in itself, a response to an enhanced stability of the molecule. Therefore, we are logically to expect that the greater the inequality in the sharing of the electron-pair by the affected atoms, the greater is the strength of their mutual chemical bond, and the larger, consequently, the expenditure of energy necessary to break it. The value of ΔE which denotes the rupturing energy of the bond must always be algebraically positive because its value depends not upon the specific electronic character of the atom nor upon the bond direction toward which the electronegativity of the electron-pair is being deflected from dead center but rather upon the *extent* of this displacement.

Thus, the polar bond Br—Cl in the polar molecule BrCl has a positive value of bond ΔE despite the fact that both of the atoms involved are highly electronegative (Group VII of periodic classification); this, just as assuredly as a bond of either of these atoms with an electropositive species on the left side of periodic classification. The shift of electrons is, in any event, always in the direction of the more electronegative species; hence, to to be depicted in this instance as $[Br^{\delta+} \overset{\times}{\times} Cl^{\delta-}]$.

Interest in bond strength or bond energy concerns not only its practical utilization to learn more about the *multi*nuclear bondings in polyatomic molecules (those considered so far are diatomic), but also an appraisal of the academic manner and justification with which electronegativity values have been established. For this latter purpose we make use of the *heats of reaction*, shown in Table 2.1, which were determined on the Pauling electronegativity scale.

Table 2.1

Heat energies of individual atoms in homonuclear molecules

$b\, \Delta$	H in H_2	F in F_2	Cl in Cl_2	Br in Br_2	I in I_2
b' (kcal/mole)/2	52.1	19.0	29.0	23.1	17.6
b'' (kcal/mole)/2	8.7×10^{-23}	3.2×10^{-23}	4.8×10^{-23}	3.9×10^{-23}	2.9×10^{-23}

Heat of reaction is the over-all quantity of heat liberated in a chemical reaction. The heats of reaction shown (at approximately 25°C) conform to our visualizations of the significance of bond energy because the heat of reaction must be equal to the difference between the strengths of the bonds newly formed in the reaction and those of the original bonds that were broken. The values employed here are affected in a minor degree from the influences of neighboring bonded atoms as well as from the numerical concessions made to room-temperature approximations. A detailed delineation of bond energies would demand their being extrapolated to the absolute temperature of zero degrees Kelvin.

The values of a'' have, in Table 2.2, been obtained merely by dividing those of a' by the Avogadro number, inasmuch as there is one bond for

Table 2.2

Homonuclear bond energies
(by Experiment)

$a\ \Delta$	H–H	F–F	Cl–Cl	Br–Br	I–I
a' kcal/mole	104.2	38.0	57.9	46.1	35.2
a'' kcal/bond	17.3×10^{-23}	6.3×10^{-23}	9.6×10^{-23}	7.7×10^{-23}	5.8×10^{-23}

each of the 6.02×10^{23} molecules present per mole of the respective diatomic substances; dimensionally, that is

$$\frac{a' \text{ kcal/mole}}{6.02 \times 10^{23} \text{ bonds/mole}} = a'' \text{ kcal/bond}.$$

To ultilize these values in the techniques of elucidating electronegativities, we first logically infer that for each of these homonuclear species each atom would contribute in precisely identical measure to the total energy of its bond with the other, as it would hardly be reasonable to assume that either of the two bonded atoms of the molecule could be any more electronegative than its chemically identical partner. Consequently, with the sharing of the electron-pair on the average equal, each bonded atom is responsible for exactly one-half of the total energy of its bond, thus yielding the values for the individual atoms of each of the bonds.

How do these heats apply to the evaluations of electronegativity pertinent to the bonding strengths of the heteronuclear hydrogen halides? Were we to infer equal sharing of the electron-pair between the hydrogen atom and each of the halogen atoms in their respective molecules, we would have to conclude, logically, that the total bond energy must be the sum of the contributions made by each atom in accordance with the applicable values of the pertinent atoms as supplied in b' or b''. This would lead to the data in Table 2.3. The numerical differences between the greater values in each instance experimentally obtained (d) and the corresponding smaller values that were calculated on a purely theoretical basis (c) imply unmistakeably the inequalities of the sharing of the electron-pairs. The order of diminishing stability of the hydrogen halides consequently follows in accordance with the values obtained in d — the progressively diminishing sequence of bond energies; that is

$$HF > HCl > HBr > HI.$$

Table 2.3

Heteronuclear bond energies;
(*Equal Sharing*)

Δ	H–F	H–Cl	H–Br	H–I
c Calculated				
c' kcal/mole	52.1 +19.0 ――― 71.1	52.1 +29.0 ――― 81.1	52.1 +23.1 ――― 75.2	52.1 +17.6 ――― 69.7
c" kcal/bond	8.7×10^{-23} $+3.2 \times 10^{-23}$ ――――― 11.9×10^{-23}	8.7×10^{-23} $+4.8 \times 10^{-23}$ ――――― 13.5×10^{-23}	8.7×10^{-23} $+3.9 \times 10^{-23}$ ――――― 12.6×10^{-23}	8.7×10^{-23} $+2.9 \times 10^{-23}$ ――――― 11.6×10^{-23}
d Experimental				
d' kcal/mole	135.0	102.1	85.9	70.4
d" kcal/bond	22.4×10^{-23}	16.9×10^{-23}	14.2×10^{-23}	11.7×10^{-23}

As, in each instance, the respective halogen atom is bonded to an atom that is common to all of the compounds — hydrogen — there is a valid comparative basis for the approximate quantitative reconciliations between the experimental bond energies and those calculated on the basis of equal sharing of the electron-pair. These mathematical reconciliations of the respective numerical differences between experimental and calculated values lead to the electronegativity assignments of the Pauling scale as well as the somewhat modified refinements of the Mulliken scale. Both sets of such arbitrary values are recorded within the table of periodic classifications (Appendix D). The Pauling scale places a convenient top of 4.0 on the most electronegative element, fluorine, and the other elements naturally fall into parallel proportionate positions with respect to it. When all are evaluated by procedures similar to those just undertaken to reconcile bond strengths with concomitant inequalities of electron-sharing, they are confirmed by the following progression of diminishing bond energies, as shown in Table 2.4.

These protracted expositions of chemical bond properties are important to insure complete understanding. In our chemistry studies we are

Table 2.4

Numerical difference between calculated and experimental bond energies; (*d-c*, from Table 2.3).

HF	HCl	HBr	HI
63.9 kcal/mole 10.5×10^{-23} kcal/bond	> 21.0 kcal/mole 3.4×10^{-23} kcal/bond	> 10.7 kcal/mole 1.6×10^{-23} kcal/bond	> 0.7 kcal/mole 0.1×10^{-23} kcal/bond

trying to comprehend the reasons for the differences in chemical and physical properties among different substances. All chemical reaction that ever happens in our familiar experimental environments involves the loss and gain, or rearrangement, of the electrons of the affected atoms. Old bonds must be broken; new ones must be formed. This is fundamental to all chemical change. The predictability of the nature and extent of the chemical behavior of species is, then, quite clearly a matter of their bond polarities which intrinsically incorporate the component aspects of atomic ionization potentials, electron affinity, electronegativity, and bond energies. These are the foundations for virtually all quantitative appraisals, even if only approximate, of the interplay and counterplay of the fundamental reaction forces that have been implied for a spontaneous interaction of atoms. For, spontaneity seeks the direction of greater stability and less energy represented by the chemical bond, a state of mutual aggregation rather than separate and independent existence.

These reaction forces that comprise the net or over-all spontaneous tendencies of substances universally to achieve states of smallest possible energy are:
(i) the attraction between the nuclei of atoms and their own valence electrons;
(ii) the attractions between the nuclei of atoms and the valence electrons of *other* atoms;
(iii) the repulsion between the electron clouds of atoms, whether such atoms be chemically identical or chemically dissimilar;
(iv) repulsion between the nuclei of atoms, whether chemically similar or chemically dissimilar.

Perhaps not immediately apparent is the need to assign reasonable formulas to chemically involved substances, if their stoichiometric reaction weights (and on occasions, their volumes also) are to be determined. The mystifying chemical phenomena of *isomerism* wherein substances of dissimilar chemical and physical properties have the same chemical formulas (molecular as well as empirical isomers) becomes satisfyingly reconciled when their geometric structures are appraised in the light of various additional attributes or properties of the chemical bonds involved; namely, *lengths* or *distances*, *angles*, and *dipole moments*.

BOND LENGTH

The length or distance of the chemical bond may be defined as the internuclear distance between any two atoms in a molecule; that is, the distance from the presumed center of positive charge of one atom to that of another with which it is bonded. It should be understood that any numerical assignments made by calculation or by experimental measurement can, at best, be

PROPERTIES OF CHEMICAL BOND

only approximate and hardly more than an average as all atoms are in a state of constant vibratory motion.

Bond lengths are conveniently expressed in angstrom units ($1\text{Å} = 10^{-8}$ cm; or $1 \text{ cm} = 10^8$ Å) and are measurable by diffraction patterns obtained when a crystal of the given substance is subject to x radiation. When crystalline material is not available the molecular spectrum of the substance may be examined to yield equally valid results.

It is plausible that the smallest amounts of energy characterizing two bonded atoms (which, as already interpreted, conform to their spontaneous directional "motivations" for greater stability) correspond to an internuclear bond distance of greatest strength. The immediate question of interest is whether an internuclear bond distance between any two atoms of given specific identity would have to be the same, regardless of the identity of the molecule in which that bond appears. Inasmuch as our earlier development has already emphasized the principle that the properties of the chemical bond are predominantly manifestations of the independent attributes of the component atoms, we should ask how bond length, as well, conforms to this principle. The measured internuclear distances of some bondings in familiar chemical compounds are shown in Table 2.5.

Table 2.5

Type and distance of some chemical bonds

oxygen-to-hydrogen		hydrogen-to-carbon		carbon-to-carbon	
in water (H_2O)	0.96Å	in methane (CH_4)	1.10Å	in diamond ($CC_4)_n$	1.54Å
in hydrogen peroxide (H_2O_2)	0.97Å	in ethane (C_2H_6)	1.10Å	in ethane (C_2H_6)	1.54Å
in formic acid (HCOOH)	0.96Å	in propane (C_3H_8)	1.10Å	in propane (C_3H_8)	1.54Å
in methanol (CH_3OH)	0.96Å	in ethylene (C_2H_4)	1.09Å	in ethanol (C_2H_5OH)	1.55Å

The virtual constancy of the bond lengths in each of these typical categories would suggest that internuclear distances are, indeed, properties of a chemical bond that is predominantly dependent on the chemical nature of the bonded atoms. We must, however, resolve the apparent contradictions in the divergence from such constancy:

Carbon-to-Carbon bonds,	Length, Å
In ethane (C_2H_6)	1.54
graphite ($CC_3)_n$	1.42
benzene (C_6H_6)	1.39
ethylene (C_2H_4)	1.33
acetylene (C_2H_2)	1.20

The progressive shortening of the carbon-to-carbon bonds exhibited in this series of compounds immediately suggests a progressive increase in the respective electron densities in the region between the nuclei of the two bonded carbon atoms of each of the molecules. This is entirely plausible in explaining a decreased bond distance, inasmuch as increased numbers of electrons in the internuclear regions would increase the electrostatic pull upon the adjoining positively charged carbon nuclei, thus drawing them closer together. If this is so, we would expect that the bond would be both shorter and stronger.

This general trend toward increase of strength with decrease in bonding distance is, at least, qualitatively followed, even though any quantitative appraisals must necessarily be crude and fraught, in any event, with the experimental ineffectualities of measurement. Nonetheless, ample justification is offered by such figures for the familiar concepts of single, double and triple bonds which relate decreases in length and increases in bond strengths to increases in internuclear electron densities. In illustration, the following formulations are entirely reasonable:

$$\begin{array}{cc} \text{H H} & \text{H H} \\ \text{H:C:C:H} & \text{H:C::C:H} \\ \text{H H} & \\ C_2H_6(\text{ethane}) & C_2H_4(\text{ethylene}) \end{array}$$

C—C $\begin{cases} \text{bond length} = 1.54 \text{ Å} \\ \text{bond energy} = 81 \text{ kcal/mole} \end{cases}$ C—C $\begin{cases} \text{bond length} = 1.33 \text{ Å} \\ \text{bond energy} = 146 \text{ kcal/mole} \end{cases}$

$$\text{H:C:::C:H}$$
$$C_2H_2(\text{acetylene})$$

C—C $\begin{cases} \text{bond length} = 1.20 \text{Å} \\ \text{bond energy} = 200 \text{ kcal/mole} \end{cases}$

Conformity, in general, with these rules is not an exclusive manifestation of C—C bonding. Thus the following approximate data are offered in further illustration:

N-to-N bonding $\begin{cases} l_{N-N} = 1.47\text{Å} \\ \Delta E_{N-N} = 37 \text{ kcal/mole} \end{cases}$ $l_{N=N} = 1.25\text{Å}$
$\Delta E_{N=N} = 125 \text{ kcal/mole}$

N-to-O bonding $\begin{cases} l_{N-O} = 1.45\text{Å} \\ E_{N-O} = 50 \text{ kcal/mole} \end{cases}$ $l_{N=O} = 1.22\text{Å}$
$\Delta E_{N=O} = 140 \text{ kcal/mole}$

$l_{N\equiv N} = 1.10\text{Å}$
$\Delta E_{N\equiv N} = 225 \text{ kcal/mole}$

$l_{N\equiv O} = 1.06\text{Å}$
$\Delta E_{N\equiv O} = 245 \text{ kcal/mole}$

with "*l*" denoting bond length, ΔE denoting bond energy, and single, double, and triple dashes representative of one, two, and three pairs of shared electrons, respectively.

Appraisals of the data supplied for bond length and bond energy pose the interesting questions of the intermediate values, between those of single and double bonds, and between those of double and triple bonds. This concerns what is generally called the *bond order;* that is, whether it be one (the single), two (the double), three (the triple), or some *fractional* "in-between" value. Were we to plot bond length against known bond orders of other species we would obtain for the C—to—C bond in *graphite* (l = 1.42Å). a *bond order* of 1.33; and, similarly, for *benzene* (l = 1.39Å), a *bond order* of 1.50. When these estimates (admittedly, highly approximate) are cautiously applied to visualizations of the structural make-up of the molecules they do lead to some interesting concepts of *resonance* — the need, that is, to utilize more than one electronic formula to portray adequately the properties of a molecule that will yet comply with the requirements of the *octet rule*.

RESONANCE FORMULATIONS

The bond orders of 1.33 and 1.50 clearly suggest the need to formulate each of the given molecules as a hybrid between a bond order of 1 and a bond order of 2. These may well constitute two or even three different formulations of electron density for the single molecule which, in collective superimposition in the mind's eye, would best represent an over-all picture of the specific species.

The multiplicity of assigned structural forms for a given substance that can have only one form is called *resonance*. The classic difficulty over the concept of resonance is the disposition to regard the molecule as oscillating from one to another of the assigned forms. The fault is not with the molecule but rather with the human inability to portray or describe this hybrid by the single structural formula that characterizes it.

An example or two will clarify what otherwise might still be confusing. The sulfur dioxide molecule (SO_2) contains 18 valence electrons, (six for the sulfur atom and six each for the two oxygen atoms). Any valid structural compliance with the octet rule requires a rigid and careful conformance likewise with the molecule's experimental properties. Pertinent to the formulation of this molecule are the observed facts that it is nonparamagnetic (no unpaired electrons present) and that the S–to–O bond is of an order intermediate between 1 (single bond) and 2 (a double bond).

The high polarity of this molecule establishes its structure as bent rather than linear. Consequently, the following structural variations,

$$\overset{\cdot\cdot}{\underset{(1)}{:\!\overset{..}{O}:\quad\quad:\!\overset{..}{O}:}}\overset{S}{}\cdot\quad\quad\text{and}\quad\quad\overset{\cdot\cdot}{\underset{(2)}{:\!\overset{..}{O}:\quad\quad:\!\overset{..}{O}:}}\overset{S}{}$$

must both be excluded. The first does not conform to the octet rule, because there are only six electrons around the sulfur atom; the second contradicts the established experimental fact that the molecule lacks the paramagnetism (weak attraction for an electromagnetic field) associated with unpaired electrons. Any attempt to satisfy octet-rule adherence (a quite arbitrary simplification) and nonparamagnetism with the following formulation,

$$\ddot{O}\!:\quad \overset{\ddot{S}}{}\quad \cdot\!\ddot{O}$$

leads to the additional perplexity that one of the S–to–O bond lengths depicted therein is of first order (single), and the other is of second order (double). This representation is in direct conflict with the experimentally-ascertained information that both S–to–O bonds are *equivalent;* that is, that although they are of the same length, the individual bonds are longer than double bonds but shorter than single bonds. Consequently, we must *imagine* the resonance picture as that of a single resonance hybrid of the two "alternate" forms,

$$\ddot{O}\!:\quad \overset{\ddot{S}}{}\quad \cdot\!\ddot{O}\!:\qquad \text{and} \qquad \ddot{O}\!:\quad \overset{\ddot{S}}{}\quad \cdot\!\ddot{O}$$

By "mental superimposition" these yield the molecular characteristics we have attempted to justify.

In similar fashion, the two contributory resonance forms of the benzene molecule, C_6H_6, established as an hexagonal ring-structure of six carbon atoms (with all six C–to–C bonds fully equivalent), that comply with the determined bond order of 1.50 (bond distance = 1.39Å), are:

Similary for the carbonate ion, CO_3^{2-}, *three* resonant forms have been established. These contribute in individually equivalent fashion to the collective representation of the experimental properties of the ion, conforming to its electronic configuration of 24 electrons, six for each of the three oxygen atoms, four for the carbon atom, and two for the ionic charge; thus

Each single dash between atoms symbolizes a shared pair of electrons.

PROPERTIES OF CHEMICAL BOND

DIPOLE MOMENT

The degree of ionic character and, hence, of the electrical asymmetry of a polar covalent molecule is quantitatively measured by its *dipole moment*, μ. This is, numerically, the value obtained by multiplying the distance, d, separating two charges of equal electrical magnitude, but of algebraically opposite sign, by the quantity, q, of such individual electrical charge; that is

$$\mu = qd.$$

That the incremental charges on the respective ends of the dipole molecule are both equal and opposite is comprehended readily enough with an awareness that the displacement of one or more electrons of the less electronegative atom of a bonded pair to a residence predominantly in the more electronegative region must inevitably leave the former with a positive charge equal in net amount to that of the negative charge now newly acquired by the latter.

As the charge upon an electron is 4.8×10^{-10} electrostatic unit (esu), and with the distance of its displacement from the less electronegative atom of the bond conveniently evaluated in angstrom units ($1\text{Å} = 10^{-8}$ cm), it is readily seen that the dipole moment μ is of magnitudes of the order 10^{-18} esu-cm. This is frequently further simplified by utilizing the unit *debye* (1×10^{-18} esu-cm; after P. J. W. Debye, 1884–1966). The displacement of the fundamental electron charge of 4.8×10^{-10} esu over a distance of one angstrom unit would then yield, in debyes (D), a dipole moment of

$$\mu = \frac{4.8 \times 10^{-10} \text{ esu} \times 1 \times 10^{-8} \text{ cm}}{1 \times 10^{-18} \text{ esu-cm /D}}$$

$$= 4.8\text{D}.$$

Dipole moment is calculated by utilizing an experimental characteristic of a substance called the *dielectric* constant. As a dipole molecule has a terminal positive and a terminal negative increment of electrical charge (as a whole unit it is net neutral) it can orient itself in an electric field (such as the oppositely charged metallic plates of an electrical condenser) with its positive end toward the negative plate and its negative end toward the positive plate. The dipole molecule can only turn; however, it cannot *migrate* to either plate inasmuch as both incremental terminal charges are of identical magnitude and, moreover, are both part of the same molecule.

The quantity of charge that can be stored upon the plates of an electrical condenser is directly proportional to the difference of potential between the plates. It follows, consequently, that if any substance placed between the plates of the condenser is capable of partially neutralizing the electrical charge upon the plates, it permits a greater storage of externally supplied electricity thereon, inasmuch as the difference of potential between the plates is increased.

If the capacitance of the condenser (C) be represented as the ratio of the amount of electrical charge upon the plates (q) to the potential difference (V) between them — that is, $C = q/V$ — then the dimensionless dielectric constant (ϵ) may likewise be represented as a ratio; namely

ϵ (of the dielectric medium)
$$= \frac{C_{\text{medium}} \,(= \text{capacitance of condenser with substance between plates})}{C_{\text{medium}} \,(= \text{condenser capacitance, vacuum})}$$

It might be expected that nonpolar molecules, such as the diatomic homonuclear H_2, Cl_2, etc., would have extremely little influence upon the capacitance of the condenser; that is, the preceding ratio would be very close to unity. The dielectric constant of H_2 gas, for example, is 1.00026. That slight deviations from unity actually do occur in such cases bespeaks the experimental observation that some distortion of electron distribution is always induced by the very nature of the electrical field itself. This induced polarity, or *polarizability*, indeed may be itself regarded as a property of the chemical bond, and of a degree quantitatively consonant with any permanent dipolarity of the bond.

Before proceeding with the mathematical delineation of dipole moment by utilizing the dielectric constant, it would be well to explore the obvious and hidden interrelationships of dipole bond and dipole molecule. A dipole molecule always has a permanent dipole bond; in fact, one or more, depending upon the complexity of the molecule. However, the reverse is by no means necessarily true; that is, a dipole bond does not necessarily make for a dipole molecule. The linear orientations of two polar bonds in the same molecule may well mutually cancel out each other's displacements of electron density. This would be true for example, in the CO_2 molecule which has two polar bonds but no permanent dipolarity as a molecule. Thus, in the pertinent representation of this molecule as

$$\overset{\times\times}{\underset{\times\times}{O}} \;\; C \;\; \overset{\times\times}{\underset{\times\times}{O}}$$

residual increments of charge δ^- $\underbrace{\delta^+\delta^+}_{2+}$ δ^-

the two energetically equivalent C-to-O bonds in this linearly symmetric molecule reveal why the molecule, with both ends negative, would have an extremely low dielectric constant. Any attempt on the part of the molecule to turn in one direction in an electric field would be immediately counteracted by an equal tendency to turn in the other. It should hardly be necessary to caution against any impetuous conclusions that linear symmetry is an invariable condition of molecular nonpolarity; or that the terminal atoms of the molecule are necessarily identical chemically. Although the ample references hitherto made to heteronuclear diatomic molecules stress the differences in electronegativity of the different terminal atoms and of the

PROPERTIES OF CHEMICAL BOND

corresponding parallel trends toward polarity and ionic character, more complex molecules require careful consideration of their precise chemical bondings and of the orientations of these bonds.

SHAPES OF MOLECULES

A few generalized situations illustrate the variabilities that have been suggested. Thus, with the letters A, B, C, and D representing different atoms, and δ representing the residual increments of charge, the *triangular planar* molecule in Figure 2.1 (a), typifies $GeCl_3$, and BF_3. Such a molecule is symmetrical and nonpolar; hence, its effective permanent dipole moment is zero. The three A-B dipole bond moments have cancelled one another.

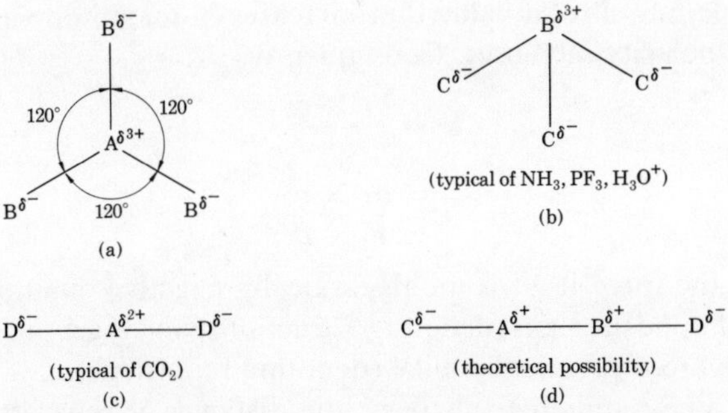

FIG. 2.1 Shapes of molecules. (a) Triangular planar. Typifies $GeCl_3$ and BF_3. (b) Trigonal pyramidal. Typifies NH_3, PF_3, H_3O^+. (c) Nonpolar linear. Typifies CO_2. (d) Polar linear. Theoretical possibility.

Additional illustrations are:

Trigonal pyramidal molecule. This is nonsymmetrical and polar; hence, a molecule with a significant dipole moment. [Fig. 2.1 (b)].

A *linear* molecule, *nonpolar:* This molecule has virtually zero dipole moment. The two A-D dipole bond moments, in effect, mutually cancel each other. [Fig. 2.1 (c)].

A *linear* molecule, *polar:* This is if the quantity of the incremental charge on the respective terminals C and D differ. Such differences, shown in Fig. 2.1 (d), would arise if the numerical difference in electronegativity of B and D is not identical with the difference in electronegativity of A and C. Consequently, C and D would differ with respect to electron density and bond dipole moments would not cancel out. The molecule as an entity would have an effective permanent dipole moment.

DERIVATION OF DIELECTRIC CONSTANT FROM COULOMB'S LAW

The electrostatics of deriving the dielectric constant by means of Coulomb's Law provides that the force of attraction (F) between two oppositely charged substances or the force of repulsion between two similarly charged substances varies in direct proportion to the numerical product of the two charges (q_1 and q_2, respectively), and inversely with the square of the distance (d) between such charges. As polar chemical particles placed between the oppositely charged plates of the condenser always lower the attractional forces measured between the plates under conditions of their inter-regional vacuum, it follows that a numerical value may be assigned for each substance which validly represents its capacity to decrease such force. This numerical assignment is the dielectric constant ϵ. It must be regarded not only as a denominator proportionality constant of value greater than unity for a polar molecule but, also, a value that increases in magnitude as the degree of molecular polarity increases. Consequently,

$$F = \frac{q_1 \times q_2}{\epsilon d^2}$$

$$\epsilon = \frac{q_1 \times q_2}{F \times d^2}.$$

Clearly, the force itself is an algebraically negative quantity when the charges are of opposite signs (denoting attractions) and algebraically positive when charges are algebraically alike (denoting repulsions).

As the force required to increase the distance of separation between two unlike charges diminishes with increase in the value of the proportionality constant (ϵ) assigned to the material medium between the charges, it follows that increasing magnitudes of dielectric constants qualitatively reflect the increasing polarity of the material media.

The applicability of Coulomb's law of electrostatics to solutions of electrolytes, in general, very convincingly interprets the efficacy of a solvent of high dielectric constant (H_2O) in assisting the process of dissolution of ionic crystals and maintaining the separations of their ions in the solution. The degrees of solubility of electrolytes in media of low dielectric constant are, as a general rule, quite negligible. Other factors, such as *lattice energy* and *hydration energy* also contribute to the energetics of solubility. These are deferred for the moment.

DERIVATION OF DIPOLE MOMENT FROM DIELECTRIC CONSTANT

An equation has been developed relating dipole moment to the dielectric constant. The expression takes into account not only the permanent dipole,

PROPERTIES OF CHEMICAL BOND

which represents a definite and characteristic orientation of the electron cloud with respect to the bonding internuclear axis of the molecule, but also the induced dipole moment or polarizability that denotes distortion of the electron cloud by the electrical field without any implication of orientation.

In the following applicable relationship,

$$\left(\frac{\epsilon - 1}{\epsilon + 2}\right)V = \frac{4\pi}{3}\left(\alpha + \frac{\mu^2}{3kT}\right)$$

"k" signifies the gas constant *per molecule* (commonly referred to as the Boltzmann constant) and having a value of 1.38×10^{-16} erg deg^{-1} molecule^{-1}, [derived as follows: $R/N =$ (gas constant per mole per degree Kelvin/Avogadro number) = (8.32×10^7 ergs per mole °K/6.02×10^{23} molecules per mole]. The polarizability of a molecule (α), obtained by measuring the molecule index of refraction of visible light is, as already noted, a response uninfluenced by the molecules' permanent dipole moment. T is absolute temperature; and "V" is the volume of molecule.

The objectives here do not require involvement with either the mathematical derivations or the practical utilization of this equation. It has been presented merely to indicate the availability of an approximate quantitative relationship between dipole moment and dielectric constant and to show how it looks.

Table 2.6 gives some values at room temperature (25°C) to demonstrate the relationships that may be derived from the preceding equation.

Table 2.6

Substance	Dielectric constant of liquid	Dipole moment, of Gaseous molecule
water, H_2O	1.85D	78.5
methyl alcohol, CH_3OH	1.70D	32.6
ethyl alcohol, C_2H_5OH	1.69D	24.3
sulfur dioxide, SO_2	1.61D	15.6
chloroform, $CHCl_3$	1.01D	4.81
benzene, C_6H_6	0.00D	2.25
carbon tetrachloride, CCl_4	0.00D	2.20

With the means established for computing dipole moments — that is, reasonably acceptable values, in the light of the experimental difficulties encountered — we may interpret some additional facets of the molecular polarity and ionic character. Let us consider in illustration the single HBr molecule. The data we use for a specific temperature, are its computed dipole moment, 0.78 debye (D), and its internuclear H-Br bond distance, 1.40Å. Utilizing the already-noted relationship,

$$\mu = q \times d$$

we may calculate q (the incremental terminal charge of the molecule) as

$$q = \frac{\mu}{d} = \frac{0.78\text{D} \times 10^{-18} \text{ esu-cm/D}}{14.0\text{Å} \times 10^{-8} \text{ cm/Å}}$$

$$= 0.56 \times 10^{-10} \text{ esu}.$$

This value is only a fractional part of the total quantity of 4.8×10^{-10} esu that constitutes the fundamental charge of the electron. The implications are obvious; we are not contending with a purely ionic bond. Such a bond would demand a transference of the full amount of this fundamental charge from one terminal of the molecule to the other. In other words, the ionic bond is to be described exactly by the relationship

$$\delta^+ = \delta^- = 4.8 \times 10^{-10} \text{ esu}.$$

As a polar covalent molecule, however, the percentage of ionic character of the HBr clearly conforms to the evaluation,

$$\frac{0.56 \times 10^{-10} \text{ esu}}{4.8 \times 10^{-10} \text{ esu}} \times 100,$$

yielding

$$\% \text{ of ionic character} = 12\% \text{ (circa)}.$$

THE INFLUENCES OF HYDROGEN BONDING

The manifest discrepancies between values obtained by the method just illustrated and those obtained using electronegativity assignments (as described, elsewhere) merely point up the significant variabilities in the approximations inherent in the different approaches, including, as well, the temperature influences. A considerable part of the difficulty in accurately relating dielectric constant to dipole moment arises from the conceived phenomenon of *hydrogen bondings* between and among individual molecules. In this concept, the pertinent molecules are presumed to be linked covalently by hydrogen atoms that act as *bridges* between small and highly electronegative atoms in the molecules, notably those of fluorine, oxygen, and nitrogen. Figure 2.2 shows a few types of hydrogen bonding.

In confirmation of Figure 2, determinations of molecular weight have revealed the existence of polymeric forms of water molecules $(H_2O)_n$ with values of n as high as 9.

The two atoms that are hydrogen-bonded need not be chemically identical. Thus, low-temperature hydrates of composition $x\text{NH}_3 \cdot y\text{H}_2\text{O}$ are known to exist, of which $\text{NH}_3 \cdot \text{H}_2\text{O}$ has been isolated experimentally. Its electron formulation responds to

$$\begin{matrix} \text{H} & \text{H} \\ \text{H:N:H:O:H}. \end{matrix}$$

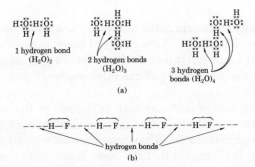

FIG. 2.2 Configuration of hydrogen bonding. (a) Water "molecules," $(H_2O)_n$. (b) Hydrogen fluoride "molecule," $(HF)_n$.

The objective of the references to hydrogen bonding here has been to interpret some of the abnormality in the relationships between the dipole moments of hydrogen-bonded species and their measured dielectric constants. The latter values are always significantly higher than those otherwise reasonably predictable from dipole moments. The average dipole moment of a hydrogen-bonded aggregate could be expected to be larger than that of a single isolated unit of it. These averages, which constitute the effective dipole moments of hydrogen-bonded species, are, moreover, quite disproportionate for the pertinent polymeric formulations. This is undoubtedly a reflection of the relative inconstancies, in general, of the number of hydrogen bridges and of the consequent values of n in the respective polymeric formulations.

What has been stated here with respect to the influence of hydrogen bonding on the dielectric constant applies also to other physical constants of the affected molecules. All figures for boiling points, freezing points, melting points, heats of vaporization, etc., for molecules of NH_3, H_2O, and HF (among others) are anomalously larger than what normally would be predicted. Thus, were we to plot for the family hydrides, HCl, HBr, and HI any of the aforementioned physical constants against the respective molecular weights, we could obtain by extrapolation the corresponding physical constant that conforms to the molecular weight of the family co-member, HF. If then, we were to compare this extrapolated constant with that obtained by actual experimental measurement, we would find the experimental value always to be much *greater* than the extrapolated value. The same is true in varying degrees of NH_3, when it is contrasted with its family co-member hydrides, PH_3, AsH_3 and SbH_3; and also with respect to H_2O in its contrasts with the co-member family hydrides H_2S, H_2Se, and H_2Te.

BOND ANGLE

If we are logically to interpret the shapes or structures of molecules that contribute to our knowledge of the chemical bond, the usefulness of dipole moments cannot be underestimated. The vectorial properties of the dipole moment permit the determining of the bond angles of complex polyatomic molecules.

The *bond angle* of a molecule may be defined as the average angle of internal intersection of the two axes that connect the nucleus of a central atom with the separate nuclei of two other atoms with which that central atom is bonded. It is stressed that the angle is necessarily an average inasmuch as all atoms are in a state of incessant vibration.

Since the net dipole moment of a molecule is the vector sum of dipole moments of the contributory individual bonds the moment of a particular bond in the molecule can be determined from the net dipole moment and the angle of the bond. The latter is obtained readily enough by spectroscopic examination of the molecular material. By way of illustration, let us seek the dipole moment of the O–H bond in the H_2O molecule. Spectroscopic evaluation indicates an O–H bond angle, θ, of 105°; and the molecule's net dipole moment is evaluated as 1.85D. The simple vector diagram shown in Figure 2.3 reveals the required moment of the O–H bond to be 1.60 debyes.

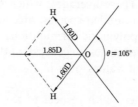

FIG. 2.3   Dipole moment of O-H bond in molecule of water. Vector diagram.

Consonant with many experimental observations, a few helpful qualitative generalizations may be made with respect to the predictability of bond angles, and of the more stable of the geometric molecular and ionic structures to which they contribute. These rules have to do not only with the numbers of electron pairs in the valence shell of the central atom that are being actually utilized in bonding (*bonding pairs*), but also with electron pairs in the valence shell that remain as yet unshared with other atoms (*lone pairs*).

Likewise to be qualitatively evaluated are the influences, in general, of electronegativity of central atoms upon the directions of the bonds, and of the latter's collective orientations in the shaping of structure. These considerations follow:

PROPERTIES OF CHEMICAL BOND

1. Electron-pairs, bonded and unbonded, tend to orient themselves into orbitals at distances as far removed from one another as possible, thus reducing to the minimum the mutual repulsions of the electrons.

2. Considered as a separate and independent effect, all other factors constant, the greater the number of lone pairs of electrons on the central atom, unshared or unbonded pairs, the greater, in general, will be the tendency of bond angles to contract.

Thus, the bond angle formed between the central atom A and the atoms B' and B'' ($\measuredangle$ B'–A–B'') is likely to be significantly larger when just one unshared electron-pair still remains on A than when two lone pairs are present there. And, likewise, with no lone electron pairs on the central atom, the bond angle could be expected to be larger still. Hence, it may be generalized:

$$[\measuredangle \text{ B}'-\text{A}-\text{B}''] > [\measuredangle \text{ B}'-\overset{\times\times}{\text{A}}-\text{B}''] > [\measuredangle \text{ B}'-\overset{\times\times}{\underset{\times\times}{\text{A}}}-\text{B}''].$$

As nonbonding electrons are not equivalent to bonding electrons in their respective orientations in the molecule as a whole, the plausible explanation that suggests itself for the increase in the angle-contracting effect, as the number of nonbonding electron pairs increases, is that nonbonding electron-pairs take up more space than bonding pairs. Consequently, this must leave less space for bonding pairs; and the geometric shapes of molecules reflect these distortions by their divergence from symmetry.

Expansion of the central atom's valence shell permits the presence of even greater numbers of lone electron-pairs on the central atom than the last of the preceding examples provided: thus,

$$\overset{\times\times}{\underset{\times\times}{\overset{\times\times}{\text{A}}\text{B}_2}}.$$

If the foregoing appraisals are to be meaningful, they should be between and among analogous molecules wherein the chemical identities of the bonded atoms are the same. It must be borne in mind that bond angles vary with the differing degrees of repulsive forces existing between the electron clouds of bonded atoms; consequently, these forces affect bond distances, or proximity to the central atom. This, manifestly, would have to be equated in terms of the respective sizes and electronegativity of the pertinent bonded atoms.

3. Considered as a separate and independent effect, all other factors constant, the smaller the electronegativity of the central atom the smaller is the bond angle.

This would certainly seem plausible enough, inasmuch as the smaller the electronegativity the greater is the bond distance between the central atom and each of the atoms with which it is bonded. Consequently, there

50   PROPERTIES OF CHEMICAL BOND

are greater opportunities for the centrally-bonded atoms to bend inward toward each other, with resultant narrowing of the angle at which their respective internuclear axes with the central atom intersect.

4. Of the aforementioned two trends with respect to bond angles — increased contraction with increased number of lone electron-pairs, and increased expansion with increased electronegativity — the former (lone-pair number) is by far the more important; that is it is qualitatively predominant.

Table 2.7 exemplifies the admittedly rough yet qualitatively useful rules that have been presented.

Table 2.7

Bond angles of hydrides of representative elements

sp^3 orbital hybrid	Periodic family		
	group IV	group V	group VI
Periodic Series 2 $2s^2, 2p^2 2p^2 2p^2$	[∡ H–C–H (in CH_4) = 109.5°]	> [∡ H–$\overset{xx}{N}$–H (in NH_3) = 107.3°]	> [∡ H–$\overset{xx}{\underset{xx}{O}}$–H (in H_2O) = 104.5°]
Periodic Series 3 $3s^2, 3p^2 3p^2 3p^2$	[∡ H–Si–H (in SiH_4) = 109.5°]	> [∡ H–$\overset{xx}{P}$–H (in PH_3) = 93.3°]	> [∡ H–$\overset{xx}{\underset{xx}{S}}$–H (in H_2S) = 92.2°]
Periodic Series 4 $4s^2, 4p^2 4p^2 4p^2$	[∡ H–Ge–H (in GeH_4) = 109.5°]	> [∡ H–$\overset{xx}{As}$–H (in AsH_3) = 91.8°]	> [∡ H–$\overset{xx}{\underset{xx}{Se}}$–H (in H_2Se) = 91.0°]
Periodic Series 5 $5s^2, 5p^2 5p^2 5p^2$	[∡ H–Sn–H (in SnH_4) = 109.5°]	> [∡ H–$\overset{xx}{Sb}$–H (in SbH_3) = 91.3°]	> [∡ H–$\overset{xx}{\underset{xx}{Te}}$–H (in H_2Te) = 89.5°]

In conformity with the trends that have been expressed, the data in Table 2.7 can be interpreted as follows:

In going from left to right in any of the periodic series (horizontal sequences) we are proceeding in the direction of increased electronegativity; hence, as a separate effect, as bond distance becomes shorter, bond angles should increase from left to right. We note, however, that as bond distance becomes shorter in going from left to right we are likewise proceeding in the direction of increased numbers of lone electron-pairs; consequently, bond angles should decrease from left to right. The sizeable progressive net diminutions of the numerical values of the bond angles in thus proceeding from left to right points up the predominance of the effect of a lone-electron pair over that of the electronegativity influence.

The family trends in the periodic groups (vertical sequences), with the exception of Periodic Family Group IV, show a definite progressive contraction of bond angles from top downwards. As we proceed downward in a family group the electronegativity of the central atom diminishes; hence, bond angles diminish for these completely analogous molecules (all being hydrides of the same general formula). Note, however, that the number of lone electron-pairs upon the central atom, as a characteristic of the entire family, remains constant. Hence, although the number of lone-electron pairs remains a significant determinant of the net numerical value of the bond angle, the qualitative direction conforms to the factor that changes — which is, the electronegativity of the central atom. Consequently, bond angles will diminish as we proceed downwards in a family group of similar compounds.

Why do the bond angles of the hydrides of Group IV (CH_4, SiH_4, GeH_4, SnH_4) remain constant despite the progressive contractions expected from their progressively diminished electronegativity as we proceed downward? The shapes of the molecules of these hydrides in Group IV are regular tetrahedra; that is, tetrahedra whose symmetry is undistorted by the greater spatial demands of nonbonding electron-pairs, which characterize the respective hydrides of Family Groups V and VI. In regular tetrahedra *all* bond angles remain constant at 109.5°. The irregular tetrahedra that structurally characterize *partial* utilization in bonding of the central atom's four available electron-pairs (Groups V and VI), and which show bond angles diminished from the regular tetrahedral angle of 109.5°, are considered later.

BOND DIRECTIONS IN MOLECULES AND IONS

In Figure 2.4 the separate atoms are represented by the letters A and B. In each instance the central atom is A, and the coordinated atom is B. There is a multiple number of B atoms in the given molecule, but it is to be understood that they are not necessarily of the same chemical identity; rather B is a purely symbolic designation for any atom that bonds with the central atom. Electrons are denoted by dots or crosses. The delineations here are restricted to nontransitional central atoms inasmuch as elaborations of transitional species are made in a separate chapter.

(1) In the *linear* orientations of bonds in AB_2, there are *two pairs of electrons, in total, in the valence shell of the bonded central atom.* See Figure 2.4.

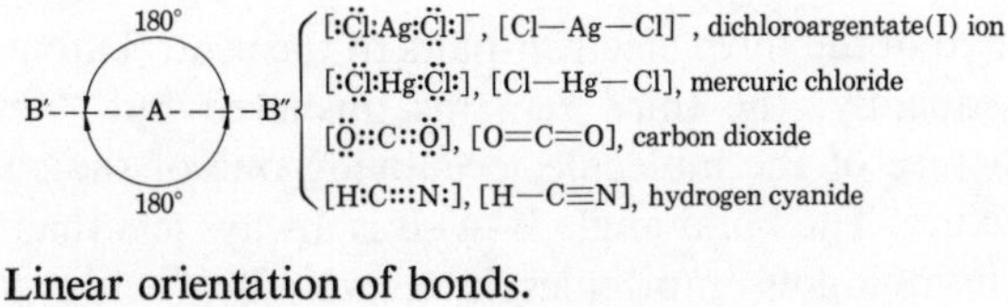

[:C̈l:Ag:C̈l:]⁻, [Cl—Ag—Cl]⁻, dichloroargentate(I) ion
[:C̈l:Hg:C̈l:], [Cl—Hg—Cl], mercuric chloride
[Ö::C::Ö], [O=C=O], carbon dioxide
[H:C:::N:], [H—C≡N], hydrogen cyanide

FIG. 2.4   Linear orientation of bonds.

In this category, both electron-pairs on the central atom are fully utilized in bondings with two other atoms; consequently, there are no

52 PROPERTIES OF CHEMICAL BOND

unused valence electrons. Bond angles of these structures are always 180° — the very farthest separation possible between two pairs of electrons. The bonds involved may be single, double, or triple.

Also included in the linear presentation are all diatomic polar and nonpolar molecules because they cannot have more than a single bond axis.

As the linear form here uses the single s and only one of the p orbitals of the valence shell of the central atom, the orbital category is called "sp hybridization."

(2) In the *triangular planar orientations* of bonds in AB_3 and in $\overset{x}{\underset{x}{}}AB_2$ (Fig. 2.5), there are *three pairs of electrons, in total, in the valence shell of the central bonded atom.*

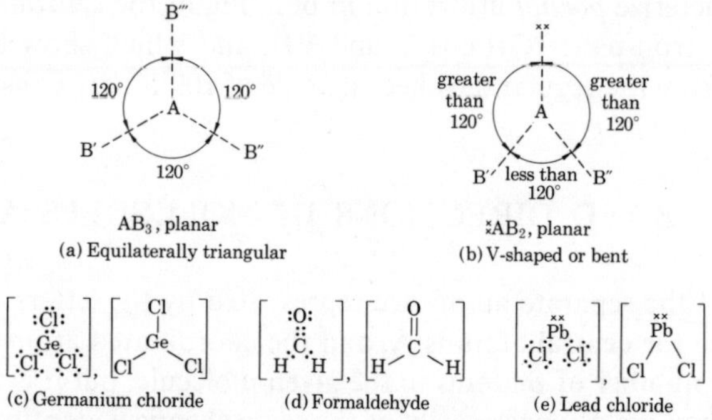

FIG. 2.5 Triangular orientations of bonds.

The two forms of two-dimensional triangular structure shown in Figure 2.5 conform to the following respective conditions:

(*a*) All three electron-pairs on the central atom are fully utilized in bondings with three other atoms. Consequently, no lone electron-pairs remain unutilized on the central atom. Bond angles herein are all 120°, representing the farthest separations mutually possible among three groups of electrons. Illustrations of this molecular type are shown in Figure 2.5, (c), (d).

(*b*) Only two of the three electron-pairs of the central atom are actually bonded. Consequently, the third remains unshared but "amalgamated" within the structure of the molecule, occupying one of the corners of the triangular structure. The bond angle B–A–B is always less than 120°, and it decreases for homologous molecules in a periodic family group as the electronegativity of the specific central atom diminishes.

Bond angles may vary here as far down as $\measuredangle$ circa 90°. An illustration of this V shaped or bent molecular type, is shown in Figure 2.5, (e). As the

PROPERTIES OF CHEMICAL BOND   53

triangular planar orientation here utilizes the single *s* and two of the three *p* orbitals of the valence shell of the central atom, the orbital category is referred to as sp^2 hybridization.

(3) The *tetrahedral* orientations of bonds in AB_4, in ${}^{\times}_{\times}AB_3$, and in ${}^{\times\times}_{\times\times}AB_2$, have *four pairs of electrons, in total, in the valence shell of the central bonded atom*.

In this category we observe three types of molecular orientations. The differentiations that are made concern the precise bond angles between electron-pairs — this time, four in number — and the extent to which they are utilized in actual bondings with other atoms. The simplest tetrahedral conformation is shown in Figure 2.6 (a), (b).

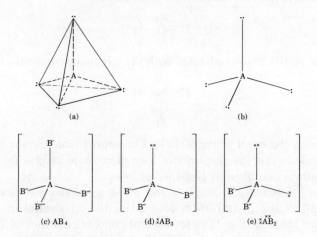

FIG. 2.6   Tetrahedral orientation of bonds.

The other examples, Fig. 2.6 (c), (d), (e), which are portrayed in the same notational manner as are the previous delineations of structure, clarify the directional nature of the pertinent chemical bonds encountered in this three-dimensional classification; thus,

(c) AB_4. When all four pairs of the electrons on the central atom are utilized in bondings, their directional orientations to the four corners of a regular tetrahedron produce B–A–B bond angles of 109° 28′. Distortions occur when the coordinated atoms are not all alike because of differing electron densities. Typical examples in this category are $SnCl_4$, $SiCl_4$, CH_4, BF_4^-, NH_4^+.

(d) ${}^{\times}_{\times}AB_3$. When three bonds are formed, leaving a single unshared pair on the central atom, the tetrahedral modification is described as a *trigonal pyramid*. The B–A–B bond angles may vary considerably from the *regular* tetrahedral angle of 109° 28′ — in some instances they decrease to values little above 90°. The variabilities are observed to be concomitant results of the influence of an unshared (lone) electron-pair in shortening the distance between two bonding electron-pairs and of the angular expansion of the distance between the bonding electron-pairs as the size of the central atom increases. Typical examples in this molecular category are NH$_3$ (∢ H–N–H = 107°) and H$_3$O$^+$ (hydronium ion).

(e) $\overset{xx}{\underset{x}{}}AB_2$. With only two bonds formed, the presence of two unshared electron-pairs on the central atom could be expected, in conformity with considerations already pondered, to contract the B–A–B angle to even less than that measured with only one unshared pair present. This expectation checks out validly when analogous molecules are appraised. Thus, the H$_2$O molecule, typical of this category, reveals an

$$\text{H—}\underset{xx}{\overset{xx}{\text{O}}}\text{—H}$$

bond angle of 104.5° as contrasted with the trigonal pyramidal

$$\text{H—}\overset{xx}{\text{N}}\text{—H}$$
$$|$$
$$\text{H}$$

molecule with the bond angle of 107°. The approximate tetrahedral shape of a molecule in this category of two lone electron-pairs plus two bonded electron-pairs is described as *V-shaped* or *bent*.

The utilization of the single *s* and all three of the *p* orbitals of the valence shell of the central atom categorizes the tetrahedral orientations herein as sp^3 hybridization. The tetrahedral bond orientations of the representative or nontransitional elements, to which we have confined present delineations, are amply augmented by those of transitional-element complexing. (Chapter 3).

(4) The **trigonal bipyramidal** orientations of bonds in AB$_5$, in $\overset{xx}{\underset{x}{}}AB_4$, and in $\overset{xx}{\underset{x}{}}AB_3$ have *five pairs of electrons, in total, in the valence shell of the central bonded atom.*

In this category we observe three stable molecular configurations. These are descriptively differentiated as *trigonal pyramid, irregular tetrahedron,* and *T-shape*; all represent the structural symmetry, shown in Figure 2.7, (*a*), (*b*).

This interesting arrangement requires some numerical definition of the bond angles formed by the five B–A–B bonds, three of which are equatorially

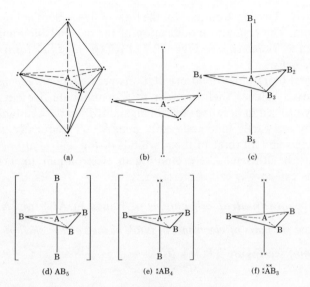

Fig. 2.7 Trigonal pyramidal orientations of bonds.

in plane with the central atom A, while the remaining two are apically directed along an axis perpendicular to the equatorial plane. Thus, for the trigonal bipyramidal structure, the foregoing description may be diagrammed as shown in Figure 2.7 (c).
In this arrangement we find

$$[\angle B_1—A—B_5] = 180°$$
$$[\angle B_1—A—B_2] = [\angle B_1—A—B_4] = [\angle B_1—A—B_3]$$
$$= [\angle B_3—A—B_3] = [\angle B_5—A—B_2] = [\angle B_5—A—B_4] = 90°$$
$$[\angle B_2—A—B_3] = [\angle B_3—A—B_4] = \angle [B_2—A—B_4] = 120°$$

In the accommodating of five pairs of electrons on the central atom the molecular orientations shown in Figure 2.6, (d), (e), (f), correspond to the numerical variabilities in the numbers of lone electron-pairs and bonded electron-pairs.

(d) AB_5. All electron pairs on the central atom are engaged in bondings; no electron-pairs can be characterized as "lone." Molecular shape is *trigonal bipyramidal*, typified by gaseous phosphorus pentachloride, PCl_5.

(e) $\overset{x}{\underset{}{\times}}AB_4$. Of the five pairs of electrons on the central atom, one remains unbonded. The resultant modification of the trigonal bipyramidal shape is called an *irregular tetrahedron*. This shape is typical of tellurium tetrachloride, $TeCl_4$.

(f) $\overset{xx}{\underset{x}{\text{X}}}AB_3$. Two electron-pairs of the total of five remain unbonded on the central atom. The resultant modification of the trigonal bipyramidal shape is described as *T-shaped*; see Figure 2.7 (f). Typical of this form is bromine trifluoride, BrF_3.

Trigonal bipramidal bond orientations necessitate an *expansion* of the central atom's valence shell in order to accommodate five electron-pairs, either fully utilized in bondings or "amalgamated" in the combinations that numerically are partly lone and partly bonded. Consequently, in addition to the single s and all three of the p orbitals of the central atom, an outer d orbital must be utilized — by promoting an electron-pair to it. Hybridization in this category is called therefore, sp^3d.

(5) The *octahedral* orientations of bonds in AB_6, in $\overset{xx}{\underset{x}{\text{X}}}AB_5$, and in $\overset{xx}{\underset{x}{\text{X}}}AB_4$, show *six pairs of electrons, in total, in the valence shell of the central bonded atom;* see Figure 2.8, (a), (b).

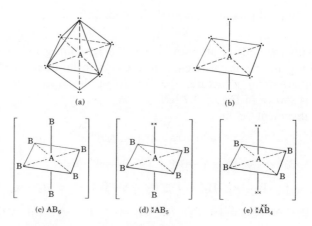

FIG. 2.8 Octahedral orientations of bonds.

The differentiations in bond orientations in this category are, again, based upon the influences that the lone electron-pairs present on the central atom exert upon the regularity of the octahedral structure:

The following variations occur, as shown in Figure 2.8, (c), (d), (e).

(c) AB_6. All electron-pairs on the central atom are fully utilized in bonding. The molecular or ionic structure is consequently that of a *regular* octahedron wherein the angle between A and any two atoms with which it is bonded is 90°. Typical of this molecular symmetry are SF_6 and AlF_6^{3-}.

(d) $\overset{x}{\text{X}}AB_5$. Five bonded electron-pairs and one lone electron-pair on the central atom yield the modification or distortion of the octahedral

PROPERTIES OF CHEMICAL BOND

molecular or ionic shape described as *square pyramid*. The molecule IF_2 is illustrative of this classification.

(e) $\overset{xx}{\underset{x}{\times}}AB_4$. Four bonded and two lone electron-pairs in the valence shell of the central atom lead, generally, to the geometric modification or *distortion* of regular octahedral symmetry called *square planar*. Typical of this classification are XeF_4, BrF_4^-, and ICl_4^-.

Clearly, expansion of the valence shell of the central atom is necessary for octahedral orientations of bonds. Such expansion is accomplished by the use in bonding of two outer d orbitals of the central atom, and it is conceived, consequently, as the promotion thereto of two bonding electron pairs from the representative element that acts as the central atom. The most stable resultant hybrid is thus constituted of six bonds, or six amalgamated lone-pair and bonded-pair combinations. Such characteristic nontransitional octahedral symmetry (or distortion thereof) is called sp^3d^2 hybridization.

ELECTRON CLOUD STRUCTURES OF BONDS

We concern ourselves herein with the types of molecular orbitals represented by σ, by π, and by δ bondings. The differentiations among these bonds are based upon the respective orientations of their clouds of electrons with respect to the common reference of the axis that, imaginatively, connects the nuclei of the atoms whose hitherto separate electron clouds have been merged into a common molecular cloud.

It must be stressed that the new orientation of electron density that results from this merger is an attribute of neither of the originally separate atoms, but is, rather, an attribute, solely and exclusively, of a completely new and independent entity, the molecule. Any visual oversimplification in picturing a molecular orbital must carefully avoid implanting a suggestion of mere superimposition of one atomic orbital upon another, purely as the mathematical net overlap. The cloud "boundary diagrams" that, figuratively, surface-encompassed about 90% of the electron density of each of the separate atoms involved in the merger have now been very significantly altered and a new boundary diagram must be drawn to describe the molecular alteration. As would logically be expected, the redistributions of individual atomic electron clouds are largely directed towards their increased concentrations in the molecular region between the atomic nuclei. This calls for a contraction of the boundary contours of the molecule as contrasted with the "overlap intermediate" and a concomitant reduction of electron density on the *far* sides or peripheral regions of the respective nuclei. This is pictured in Figure 2.9, (a), wherein the heavy dots denote the nuclei.

58 PROPERTIES OF CHEMICAL BOND

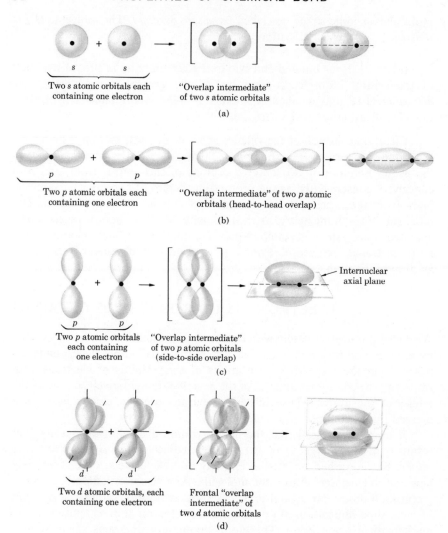

FIG. 2.9 Bonding. (a) Two-electron molecular orbital formed from two *s* atomic orbitals. A *sigma* bond: no internuclear axial plane of zero cloud density. (b) Two-electron molecular orbital formed from two two-lobed *p* atomic orbitals. A *sigma* bond: no internuclear axial plane of zero cloud density. (c) Two-electron, two-lobed molecular orbital formed from two *p* atomic orbitals by side-to-side overlap. A *pi* bond: one internuclear plane of zero cloud density. (d) Two-electron, four-lobed molecular orbital formed from two four-lobed *d* atomic orbitals. A *delta* bond: two internuclear axial planes of zero cloud density.

The increased stability of the molecular orbital, as contrasted with each otherwise-isolated atomic orbital contributing to its make-up, reflects the diminished energy requirement of the molecular system. Remember that the proximity of approach of atoms to one another in the process of merger is dictated by the equilibrium balance between the mutual attractional pull of the positively-charged nucleus of each atom upon the electron clouds of the other, and the mutual repelling effects of the electron clouds of each. Consequently, the internuclear distance represents the equilibrium between attractive and repulsive forces. The equilibrium bond distance and equilibrium bond energy both are at the minimum at the point of greatest molecular stability. Since, by convention, we accept that the energy of each isolated atom is zero with respect to the molecular system, it follows that the maximum stability of a molecular system under any given set of experimental conditions is numerically defined by the shortest bond distance that the system can achieve and also by the highest negative value of bond energy likewise attainable. That the latter is necessarily negative is the result of starting with *zero* system-energy.

Aside from the common limitation of just two electrons to each molecular orbital — which likewise characterize the forthcoming diagrams of pi and delta molecular orbitals and their bonds — the molecular feature upon which classifying distinctions rest is the extent, if any, to which the molecular internuclear axis is included within the domain of negative electrification, the electron cloud. Hence, the distinctions of sigma,- pi-, and delta-bondings concern the number of Cartesian symmetry planes of *zero* cloud density that encompass the internuclear axis.

There is always at least one internuclear axial plane, although the cloud density of this need not be, necessarily, zero. A three-dimensional visualization of the sigma molecular orbital [see Fig. 2.9, (*a*)] against a background of Cartesian coordinates reveals that no plane of zero cloud density can encompass the internuclear axis. This is the characteristic form of a sigma molecular orbital, although sigma-bonding need by no means be restricted solely to the equilibrium interactions of *s* atomic orbitals, despite the obvious parallelism in connotation between word and letter. Thus, using the double-lobed *p* atomic orbitals we may also derive the identical characteristic of sigma-bonding; that is, no internuclear axial plane of zero cloud density.

Further reflection upon the manner in which these two *p* atomic orbitals have met in "head-to-head" overlap [Fig. 2.9, (*b*)] suggests still another way in which they may possibly be joined. This alternate method, a "side-to-side" overlap, is shown in Figure 2.9, (*c*).

Wherein lies the distinction between "head-to-head" overlap of the sigma (σ) bond, and the "side-to-side" overlap, designated as the *pi* (π) bond? It is to be found, again, in the number of planes of zero concentration or cloud density of electrons that include the internuclear axis. In pi-bonding, as just illustrated, there is only one internuclear axial plane of zero cloud density. In sigma-bonding there was *none*, nor could there ever be any at all, regardless of whether s bonds with s, or p with p, or s with p. It is this feature, then, that serves always to distinguish between sigma- and pi-bonding.

A somewhat more challenging test of three-dimensional visualization is presented in concepts of d-orbital bonding. The four-lobed d atomic orbital may unite not only with an s or a p atomic orbital, to form a sigma or a pi-bond, but also with a p or another d orbital, to form a four-lobed *delta* molecular orbital. The characteristic of the delta molecular orbital or bond is that it provides two planes of zero cloud density containing the internuclear axis. This is shown pictorially in Fig. 2.9 (d).

It may be stated that where covalence, in any of its degrees, contributes to the fundamental structural shape of a species it is the sigma-bonds that exert by far the greatest influence on it, together with the unshared electron pairs (as already described). Normally, being the strongest of the different types of bonds — a reflection of its more effective overlap as compared with that in the pi- or delta-bond — the sigma bond is the backbone of the skeletal structure of the molecule. The contributions of the progressively weaker pi- and delta-bonds to over-all molecular structure are negligible, primarily being limited to their effects of shortening bond distances and of slightly modifying bond angles. If sigma bonds be regarded as the determinants of fundamental molecular shape and symmetry, then pi- and delta-bonds may plausibly be characterized as the geometric determinants of unsymmetrical *distortions*.

The presence of pi- and delta-bonds in substances are, however, logically to be related to their increased chemical reactivity as compared to analogous species with greater numbers of sigma-bonds. This is more readily seen in the more active chemical behavior of members of the alkene homologous series of hydrocarbons than in the alkanes. Alkenes contain multiple electron-pair bonds (specifically, carbon-to-carbon double bonds), of which one is a strong sigma-bond and the other(s) weak pi-bond(s) of σ–π bonding. The carbon-to-carbon bonds of the alkanes, on the other-hand, are all single sigma-bonds. Thus, the readiness with which the two-carbon alkene, *ethylene*,

$$\begin{array}{c} \text{H} \quad \text{H} \\ | \quad\; | \\ \text{H—C}=\text{C—H} \end{array}$$

is converted to the two-carbon alkane, *ethane*,

$$\begin{array}{c} \text{H} \quad \text{H} \\ | \quad\; | \\ \text{H—C—C—H,} \\ | \quad\; | \\ \text{H} \quad \text{H} \end{array}$$

is validly interpreted as opening the σ–π double bond of the former and redistributing it as three single sigma-bonds, one carbon-to-carbon and two carbon-to-hydrogen. Lest the transposition of the *two* electron-pairs of the double bond (four electrons) into the *three* electron-pairs of the *three* newly-formed single bonds (six electrons) seem perplexing, remember that the two newly-incorporated hydrogen atoms supply one electron apiece.

Similarly, the ready disruption of the triple bond between the two carbon atoms of acetylene (ethyne), H—C≡C—H, the first member of the alkyne homologous series of hydrocarbons may be visualized as the opening of this triple bond (constituting one strong sigma- and two weak pi-bonds, σ–2π), and its reconstruction progressively into the σ–π carbon-to-carbon double bond of C_2H_4 (ethylene) and the sole σ carbon-to-carbon single bond of C_2H_6 (ethane).

THERMODYNAMIC CYCLES IN IONIC BONDING

Our earlier deliberations of the energetics of chemical bonding have concerned the interactions of hitherto isolated atoms — atoms, that is, presumed to be initially independent of each other, as is conceivable in the highly excited states of their high-temperature gas phases. Such a qualitative simplification, although justifiable in establishing a necessary basis for theoretical comprehension, is hardly in accord with the everyday realities of normal chemical reaction. Except in rare circumstances, chemical reactions are interactions among tremendous numbers of aggregated atoms, which necessarily involve first, their several and individual mutual separations from one another's electrical spheres of influence, and then altered reunions to form the likewise tremendous numbers of physical aggregates that are characteristic of the products.

In illustration, the preparation of solid sodium chloride by direct union of the elements (there are more practicable ways to make it) would not be conducted by reacting sodium vapor with chlorine gas at temperatures for which the presence of independent single atoms of sodium and of chlorine may be reconciled. Indeed, the direct union of these elements as a feasible procedure would require the addition of solid metallic sodium to chlorine gas at temperatures that would demand consideration of both the crystal lattice structure of the metallic sodium and the molecular constitution of the chlorine gas as profoundly important qualitative and quantitative factors in the over-all energetics of formation. The lattice bonds of the sodium metal must be broken, the molecules of chlorine must be dissociated into atoms, and the bonds of sodium chloride, Na^+Cl^-, must be elaborated into the giant molecular three-dimensional structure of the salt wherein each sodium ion resides within a sphere of six chlorine ions and each chlorine ion, in turn, resides within a sphere of six sodium ions. The lattice structure is depicted in Figure 2.10.

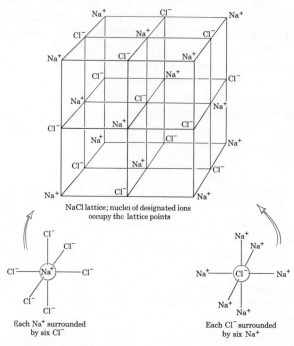

FIG. 2.10 Lattice structure of NaCl.

THE BORN–HABER CYCLE

The apparent complications of deducing the energy requirements for the formation of an ionic solid are resolved by a form of chemical bookkeeping called a Born–Haber cycle. This is an analysis based upon a strict mathematical accounting of all energy gains and energy losses occurring in a closed system. In such a system the total energy change involved in going from and returning to any step of the cyclic process must clearly be zero. This is simply an application of the first law of thermodynamics — essentially, the principle of conservation of energy. Consequently, we are in a position to evaluate any step in the cycle when other steps are known. We illustrate by seeking the net energy change of the over-all reaction to which we have just referred, all substances being in their conventional states at room temperature; namely

$$Na^0_{(s)} + \tfrac{1}{2} Cl_{2(g)} \rightarrow NaCl_{(s)}.$$

Since the heat of formation of one mole of the salt (ΔH_f) is itself a net value representing the algebraic sum of component heat energies, we may elucidate this net value from the several intermediate processes that contribute to it (Hess's law. This is a sequential analysis involving different specific amounts of heat, symbolized and defined as follows:

S — *sublimation energy* or energy that must be absorbed to vaporize the sodium metal (to the gaseous state).

I — *ionization energy* or energy that must be absorbed by the gaseous sodium atoms in order to convert them, by electron-removal, to gaseous sodium ions.

D — *bond dissociation energy* or the energy that must be absorbed in order to dissociate the gaseous chlorine molecules into gaseous chlorine atoms.

A — *Electron affinity*, or energy that must be liberated in the formation of gaseous chlorine ions from the gaseous chlorine atoms.

U — *Lattice energy of union*, or energy that must be liberated when the gaseous sodium ions join with the gaseous chlorine ions, and the subsequent condensation of these gaseous ion-pairs into the crystal lattice structure of the ionic solid, NaCl.

This lattice energy of union is calculated by applying Coulomb's law of electrostatics which, for the purpose here, takes the form,

$$U = -Me^2/r^0$$

wherein M is the *Madelung constant* (a value numerically dependent upon the mutual geometric orientations of the anions and cations that characterize the structural type of the salt crystal); e is the value of a unit electrical charge (4.8×10^{-10} esu); and r^0 is the equilibrium bond distance between ions (the distance at which potential energy is at a minimum).

Actually, what we derive here is not crystal lattice energy itself (i.e., energy liberated when gaseous ions unite to form solid salt), but, rather, only the coulombic attractions between ions being regarded as "point" charges without regard for the highly significant mutual repulsions that result from their surrounding electron clouds. Consequently, the true lattice energy is found by correcting, by numerical subtraction, the calculable negative attraction that constitutes repulsion. This, in effect, make the actual magnitude of U less negative than that computed by Coulomb's law, without consideration of such repulsion.

The five separate processes that we have defined, three endothermic (ΔH, positive) and two exothermic (ΔH, negative), yield, upon algebraic addition of their respective ΔH values, the net equation and the net ΔH of the reaction sought—as shown in Table 2.8.

The net conversion is, thus, exothermic; and formation of one mole of solid NaCl from its elements necessitates the liberation of 97.7 kilocalories of heat.

Table 2.8

Energetics of formation (f) of solid NaCl from its elements
(Room Temperature)

nature of energy change	chemical equation	heat of change
S, sublimation	$Na^0_{(s)} \rightarrow Na^0_{(g)}$	$\Delta H_S = +26.0$ kcal
I, ionization	$Na^0_{(g)} \rightarrow Na^+_{(g)} + 1e^-$	$\Delta H_I = +118.4$ kcal
$\frac{1}{2}D$, bond dissociation	$\frac{1}{2}Cl_{2(g)} \rightarrow Cl^0_{(g)}$	$\frac{1}{2}\Delta H_D = +28.6$ kcal
A, electron affinity	$Cl^0_{(g)} + 1e^- \rightarrow Cl^-_{(g)}$	$\Delta H_A = -86.5$ kcal
U, lattice	$Na^+_{(g)} + Cl^-_{(g)} \rightarrow [Na^+Cl^-]_{(g)}$ $[Na^+Cl^-]_{(g)} \rightarrow NaCl_{(s)}$	$\Delta H_U = -184.2$ kcal
f, net energy of formation	$Na^0_{(s)} + \frac{1}{2}Cl_{2(g)} \rightarrow NaCl_{(s)}$	$\Delta H_f = -97.7$ kcal

The conforming arithmetical expressions,

$$f = S + I + \tfrac{1}{2}D + A + U$$

and

$$\Delta H_f = \Delta H_S + \Delta H_I + \tfrac{1}{2}\Delta H_D + \Delta H_A + \Delta H_U$$

may be utilized in the similar stepwise energy formulations for all other conversions to alkali halides from atoms of the respective metallic solid and the gaseous halogen molecules. For these, the required intermediate cyclic ΔH values are readily measured experimentally, although the electron affinity values will be the least certain. For this reason, the Born–Haber cycle may be preferentially employed to evaluate approximations of electron affinities from the far more reliable values obtainable for the other energy terms present.

Extreme care must always be exercised with the numerical substitutions required by the equation. The appearance of negative algebraic signs in such obvious rearrangements as

$$-U = S + I + \tfrac{1}{2}D + A - f$$

sometimes beguile the unwary from the necessity of substituting negative numerical heat quantities for ΔH wherever the process is exothermic (heat evolved). These are our compulsory reconciliations with established definition that all ΔH values be quantities of heat *absorbed*.

Illustratively, the ionization energy in the expression,

$$\Delta H_I = \Delta H_f - \Delta H_S - \tfrac{1}{2}\Delta H_D - \Delta H_A - \Delta H_U$$

will validly reappear as the numerical quantity utilized in the preceding calculation when substitutions for the pertinent terms are followed in accordance with this definition, as follows:

$$\Delta H_I = -97.7 - 26.0 - 28.6 - (-86.5) - (-184.2)$$
$$= -97.7 - 26.0 - 28.6 + 86.5 + 184.2$$
$$= +118.4 \text{ kcal (as before).}$$

PROPERTIES OF CHEMICAL BOND

Properly interpreted, such thermodynamic series reveal

1. the relative magnitudes of ionic crystal lattice energy, U, are manifestations of the typically strong electrostatic forces that bind ionic crystals together and which account, consequently, for their generally high melting points and degrees of hardness.

2. the stabilities of the crystalline ionic solids are manifestations of not merely their own properties, but also of the bonding strengths of the halogen molecules and of the latticed solid alkali metal ions. The latter, in occupying the lattice points of their respective metallic structures, must be regarded as being immersed in and, consequently, bonded to the mobile cloud of their valence electrons which they have cooperatively funded within the crystal as a whole.

3. the alternate paths possible for the formation of ionic solids from their respective elements. Thus, for the formation of an alkali halide according to the following:

$$M^0_{(s)} + \tfrac{1}{2}X_{2(g)} \rightarrow MX_{(s)}$$

wherein M^0 denotes any alkali metal (Li^0, Na^0, K^0, Rb^0, Cs^0), and X_2 any diatomic halogen molecule (F_2, Cl_2, Br_2, I_2), the Born–Haber cycle may be diagrammed to reveal the two different paths possible to the formation of the salt.

Thus, in the thermodynamic cycle, shown in Figure 2.11, we observe not only the direct one-step route to the formation of solid MX by reaction between the elements in their normal states, but also the purely theoretical multiple-step process that we have just elaborated. Cycle requirements of energy remain identical for the formation of the ionic solid, whether reaction proceeds in the one step or through the several sequences.

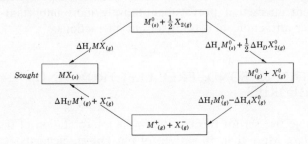

FIG. 2.11 Thermodynamic cycle.

Thermodynamic cycles such as the one just pictured may be used to interpret other trends in the energetics of ionic reaction, notably those of *oxidation potentials*. Moreover, such cycles may validly be applied to the altered physical states conforming to variable conditions of temperature and environmental media. Had temperatures above the boiling point of the pure alkali metal been specified, we would have had to exclude the vaporization step (ΔH_s of $M^0_{(s)}$) from our hypothetical breakdown of the multiple steps of the cycle. Were the temperature also above the boiling point of the alkali halide, clearly the thermal consideration of the condensation of the gaseous ion-pairs [M^+X^-] into the crystal lattice structure of $MX_{(s)}$ would certainly have had to be omitted. In this event, the sole remaining component of the evaluation of ΔH_U that concerns the union of species would have been the association of the gaseous alkali metal ion with the gaseous halide ion to form the gaseous ion-pair $[M^+X^-]_{(g)}$.

If on the other hand, the environmental conditions were those of *aqueous solution*, we would have had to introduce additional steps into the cycle, to portray the hydration of the ions. The energetics of the over-all cyclic process of forming salt in solution would then lead to evaluations significantly different from those of crystal lattice formation, inasmuch as solvation energy would have to be considered. As energies of solvation are over-all exothermal quantities, their ΔH values are algebraically negative.

To be exact, we would have to regard hydration energy (solvation energy in aqueous medium) as representing something more than just the liberation of energy due to the attraction of dipole water molecules to ions of the solute. Although energy is necessarily liberated in the ion-dipole attraction constituting the solute-solvent interactions of hydration, it is also true that energy must be expended to separate the solvent molecules from one another in order to accommodate the solute within the solvent. As the exothermal process cannot occur without a concomitant endothermal reaction, the inevitable difficulty of experimentally distinguishing the individual energy involvements of each from the other logically invites acceptance of hydration energetics as an over-all net effect. Inasmuch as $\Delta H_{hydration}$ is numerically negative (an exothermic net), the ion-dipole attractions of solute-solvent interaction prove quantitatively more important than the dipole-dipole attractions of solvent-solvent interactions.

THE METALLIC BOND; FREE-ELECTRON AND BAND THEORIES

The final objective set for this chapter is the interpretation of the metallic bond in *solids*. More than 70 percent of our known elements occur freely as homoatomic associations characterized by metallic bonding. It would be quite unimaginative, therefore, to exclude such species from a reasonable degree of exploration.

Perhaps the most familiar of the physical characteristics possessed of metals that serve to differentiate them from nonmetals is their high degree of both thermal and electrical conductivity. These properties, suggestive of movements of particles internal to the metal-atom lattice itself, as well as the malleability and ductility of most metals, (suggestive of flexible bondings) actually belie the notable strength of metals and the tightness with which their atoms are packed in the crystal.

In all metals, each atom always has a very large number of neighbors, all either equally distant from it, or very nearly so. Such a condition quite generally entails the very closest possible packing of the atoms. As assumed spheres of equal sizes, this closest packing would conform to the intimate contact of each interior atom of the crystal with twelve nearest neighbors. This metallic crystal form, called *cubic close-packing*, describes a *coordination number* of twelve. A variation of this, the *hexagonal close-packing*, likewise describes a coordination number of twelve. In another form, the *body-centered cubic lattice*, each interior reference atom is centrally oriented to eight nearest neighbors (conceived as occupying the eight corners of its own cube), and to six other neighbors only slightly farther away (conceived as the centrally oriented atoms of six adjacent cubes in respective individual contact with the six separate faces of the cube of the reference atom). This body-centered cubic lattice typifies a metallic coordination number of eight.

The approach to bonding to which we have been conditioned by the theory of valence bonds is to reconcile the number of possible principal bonds with the number of valence electrons in the electronic configuration of the pertinent atom. Yet, Al^0, with three valence electrons, should normally form only three bonds; Ca^0, with two valence electrons, should form only two bonds; and Ag^0, with only one valence electron, should form just one bond. Nonetheless, all three of these metals are cubic-close-packed in their crystal structures, and an atom of each, consequently, may be presumed to have formed twelve bonds with its twelve nearest neighbors; that is, each reveals a coordination number of twelve. If we insist upon reconciling valence with coordination number, we must assume that each of the metallic atoms we have mentioned is sharing its valence electron(s) *fractionally* with its twelve nearest neighbors. Consequently, each Al^0 atom must share *six* electrons in total with these near neighbors; for, each Al^0 has three electrons of its own and has received, additionally, a charge equivalent to three-twelfths of an electron from each of its twelve neighbors.

Similarly, it is assumed that each Ca^0 atom must share *four* electrons in total; two of which it already has, plus an equivalent of two-twelfths from each of its twelve neighbors. Likewise, the Ag^0 atom shares *two* electrons in total; one that it already has, plus one-twelfth from each of its twelve neighbors. As an observation that is mathematically justified, then, it would appear that the total maximum number, both of electrons that can be shared by the atom and, hence, also of the number of orbitals available to contain them, may be less but not greater than twice the number of valence electrons supplied by the pertinent atom.

The appraisals provided lead not only to the foundations for the existing theories of metallic bonding, but also to the electronically fundamental distinctions that exist between the atoms of metals and nonmetals. The metal has a larger number of valence orbitals than of valence electrons; consequently, vacancies are always available in metallic orbitals for the accommodating of many additional electrons. This clearly accounts for the high coordination numbers of metals. On the other hand, the typical nonmetallic atom has as many valence electrons as, and frequently more than, there are orbital vacancies; hence, the very limited coordination capacities of nonmetals are plausibly reconciled.

FREE-ELECTRON THEORY OF METALLIC BONDING

With a plethora of orbital vacancies for the accommodating of valence electrons, and a serious deficiency of such electrons for the maximum filling of such orbitals, it hardly seems warranted to contend that the valence electrons of each metallic atom would find a climate more favorable for their stability by remaining localized on the respective atom. Indeed, with the deficiency of electron density represented by the absence of an adequate layer of negative charge around each respective nucleus, it would appear entirely logical to suppose that the nucleus would seek and attract to itself the valence electrons of atoms in its vicinity. This is the point of view held in the ionic theory of metallic bonding. The *free-electron* theory of metallic bonding is that instead of remaining localized upon specific atoms and being identified with them, the valence electrons of each atom become *nonlocalized* and freely move about in a diffuse uniform electrical charge that pervades the entire metallic lattice. They are thus presumed to occupy, with relatively equal stability on the average and without constancy of tenancy, the vacancies in *all* available oribtals.

In this free-electron concept we have the simple electrostatic description of metallic bonding. As the lattice points of the metallic crystal are now occupied not by neutral atoms, but rather, by its positively charged ions, the movable current of negative electrification at once becomes the focal point of common interest for all the positive ions. All these individual ions at the lattice points tend to maintain their over-all electrical neutrality by resisting detachments from their roaming valence orbital electrons. The strength of bondings will have to be evaluated in terms of specific charges upon the positive ions, their radii, the numbers of valence electrons, and the types of orbitals contributory to the free electrons in the spaces between the ions of the crystal.

Any adopted concept that bonding must be a "personal" matter between two specific atoms can be refuted, however, in this *ionic* model. As we have conjectured, many adjacent atoms remain in mutual association solely as a result of the common and competitive attractions of their nuclei or positive ions for their nonlocalized valence electrons. The free-electron concept of the metallic bond has the virtue, not only of simplicity in plausibly accounting qualitatively for the notably high electrical conductivity of most metals, but also of not posing any conflicts with the reconciling of their high coordination numbers with their limited numbers of valence orbital electrons.

Certain questions of quantitative aspect are still unanswered. For example, why does the resistance to electrical conductance of some metals linearly increase as their temperatures are raised, whereas for others, possessed of relatively high resistances to begin with, similar elevations of temperature result in exponentially decreased resistance? And, likewise, why do some metals show virtually no conductance in the solid state and must therefore be classified as "insulators?" Obviously, something more is required in the way of visualization to bridge the theoretical gap between these extremes.

VALENCE THEORY OF METALLIC BONDING

The second theory of metallic bonding starts with the premise that all electrons of all atoms in the metallic crystal belong to the crystal as a whole and that they are distributed in groups or *bands* of discrete and quantized energy levels.

Each of these individual energy levels can hold two electrons (Pauli Exclusion Principle) but they may become so very thinly spaced when the number of atoms in the crystal becomes very large that, in effect, their energy sequences may amount very nearly to a continuum of energy. In any event, each band (starting with that of very lowest energy) is separated from the next band of energy levels by a "forbidden zone" wherein electrons may not "dwell." Inasmuch as we logically assign to the lower energy levels those electrons that do not contribute to the electrical and thermal conductance of the metal, we need concern ourselves only with the valence orbitals and valence electrons of the highest energies indigenous to the *valence band*. This band must contain more energy levels than electrons able to fill them. As the lower bands are filled with their quotas of nonvalence electrons, the number of energy levels available must be just sufficient to accommodate all such electrons.

We create now, for the composite picture of our interpretations of the conductance properties of metals, the penultimate *conduction band* which is higher in energy than the valence band. Electrons able to jump the energy gap that separates the valence band from the conduction band — the forbidden zone — thereupon enter the conduction band and, on the application of an external electrical field, have the energy required to move freely throughout the entire crystal as nonlocalized electrical charges.

Explanations with respect to conduction, semiconduction, and insulation become apparent in a study of Figure 2.12.

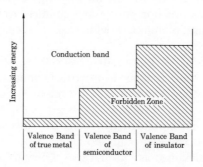

FIG. 2.12 Valence and conduction bands of metals and insulators.

(a) *True metal.* The *true metal* is characterized by excellent conductance, but shows increasing resistance as temperature rises. Here, the energy gap representing the forbidden zone between the valence band and the conduction band must be presumed to be so narrow that, under all normal circumstances, it is easily bridged by valence electrons from the valence band with their ensuing nonlocalizations throughout the crystal. It must be presumed that the forbidden zone of the true metal, always, is sufficiently narrow to permit partial filling of the conductance band with valence electrons under all circumstances, and that the high mobility of these electrons as nonlocalized entities imparts the characteristic thermal and electrical conductance to the true metallic crystal.

Why does conductivity of such a crystal diminish upon elevation of the temperature? The answer most probably lies in the increased vibratory motion of the positive ions at the lattice points of the crystal. It is certainly logical to surmise that with increasing amplitude of such vibration there will be greater interference with the free movement of the electrons in the conduction band because the mean-free paths of these electrons are diminished by collisions with the lattice structure.

(b) *Semiconductor.* The *semiconductor* is characterized by relatively poor conductance at ordinary temperatures but by *diminished* resistance to conductance as temperature increases. Note that the forbidden zone for the semiconductor is wider than that of the true metal. Under normal circumstances no electrons can bridge the energy gap or barrier and enter the conduction band; hence, no significant conduction ordinarily occurs when an electrical field is impressed upon the crystal. Under the excitation of sufficiently increased temperature, however, electrons from the valence band can bridge the gap and, as nonlocalized entities, impart highly appreciable conductance properties to the crystal.

Such is the conjectured nature of the semiconductor as contrasted with that of the true metal. How do we explain the semiconductor's increase in conductivity upon elevation of temperature? This contrasts with the behavior of the true metal. Clearly, to conform to our given picture, the transfer of electrons from the valence band of the semiconductor to its conduction band must be *promoted* with temperature increase. The explanation would have to incorporate the balance equilibrium between the increase in the numbers of free and nonlocalized electrons in the conduction band, and the decrease in their mobility with increased vibratory motion of the lattice-fixed ions.

In the semiconductor, apparently, the construction of the crystal is such as to favor the temperature factor of increased conduction band population over impairments of the mobility of the nonlocalized electrons by lattice-ion vibrations. In the true metal, the reverse situation holds; that is, mobility diminishes at a rate greater than the electron population of the conduction band.

(*c*) *Insulator*. The *insulator* should be just as readily interpreted. Here, the "forbidden zone" between the valence band and the conduction band is comparatively so wide that under the same conditions of temperature increase in which a semiconductor may be created, no electrons from the valence band of the insulator are of sufficient energy to bridge the energy barrier and gain access to the conduction band.

The band theory represents interpretations in degree rather than intrinsically. Indeed, if the temperature is high enough even the insulator may become a semiconductor. The practical choice of conductor, semiconductor, or insulator must clearly be evaluated with respect to the temperature conditions of its respective employment.

Interestingly enough, the deliberate additions of small amounts of some other metal as an impurity in an otherwise pure insulator may well make the latter a semiconductor (termed an *impurity* semiconductor). It is conjectured that the impurity atoms established separate and discrete energy levels of their own, unmixed by overlap with those of the atoms of the insulator, within the original forbidden zone of the intrinsic insulator. This, would create more favorable conditions for the separation of additional electrons either directly from the energy levels of the insulator atoms or from the energy levels of the adulterant or impurity atoms themselves.

Thus, two ways of promoting charge flow may be visualized in impurity semiconduction — depending upon the relative numbers of valence electrons of the impurity atoms as compared to those of the insulator atoms, and also upon the relative proximity of the impurity energy levels to the insulator's valence band or conduction band.

It is found, for example, that the addition of traces of boron, gallium, or indium (Periodic Group III; possess three valence electrons to the atom) to either pure germanium metal or pure silicon (Periodic Group IV; four valence electrons to the atom) converts either of the latter from an insulator to a useful semiconductor. In this instance, it is conjectured that empty energy levels have been established by the impurity within the forbidden zone of the germanium or silicon insulator, close enough to the respective valence band to permit, under thermal excitation, a ready departure of electrons from its energy levels to the empty levels of the impurity atoms. The vacancies thus created in the valence band of the insulator are to be likened to positive *holes* which, under the influence of an electrical field, are presumed to *flow* and constitute the actual carriers of the electric charge.

The requirements for this type of impurity semiconduction (positive-*hole flow*) are therefore that:

(a) the impurity atoms possess fewer valence electrons than have the insulator atoms; and

(b) the *empty* energy levels established by the impurity atoms within the forbidden zone of the erstwhile insulator be much closer to the latter's valence band than to its conduction band.

The second type of impurity semiconduction would be in contrast to the preceding situation. Energy levels would be established by impurity atoms within the insulator's forbidden zone, under conditions where

(a) the impurity atoms have greater numbers of valence electrons than have the insulator atoms.

(b) *filled* impurity-atom levels of energy are much closer to the insulator's conduction band than to its valence band.

The arrangement, here, invites the assumption that electrons in the filled energy levels of the impurity atoms find the conduction band of the insulator readily accessible to them under thermal excitation. With the energy barrier to their transitions so much less than that confronting the insulator's valence band electrons, it must be logically assumed that valence electrons of the impurity atoms, rather than of the insulator atoms, will successfully negotiate the forbidden zone and enter the conduction band of the erstwhile insulator. Consequently, under the influence of an externally applied electrical potential, the pertinent charge carriers of electricity in this type of impurity semiconduction are impurity-atom electrons which, having "invaded" the insulator's conduction band, have become non-localized throughout its crystal. The conversion to semiconduction of pure silicon (Periodic Group IV; insulator with four valence electrons to the atom) by treatment with minute amounts of arsenic (Periodic Group V; five valence electrons to the atom) is presumed to occur via this theoretical mechanism.

EXERCISES

(See note for Exercises of Chapter 1.)

1. Many exceptions have been noted to the rule that elements beyond hydrogen stabilize themselves in chemical reaction by surrounding themselves with an octet of electrons. Such violations of the octet-rule are conspicuous in the compounds ICl_3, SCl_4, and PCl_5 wherein the iodine, sulfur, and phosphorus atoms are each surrounded by ten electrons. In SCl_6, the valence shell has apparently expanded to accommodate twelve electrons. Noteworthy, too, are the now-numerous unexpected compounds of the "inert" noble-gas elements — such as XeF_2, XeF_4, XeF_6, $XeOF_2$, $XeOF_4$, XeO_2F_2, XeO_3, $XePtF_6$, KrF_2, etc. — wherein, despite the already-achieved octet of the noble-gas, its further bonding can be promoted.

The conjectures of the *expanded-octet* theory may be illustrated by using PCl_5 as an example. Since the valence electrons of the isolated phosphorus atom conform to the orbital description

$$3s^2\ 3p^1\ 3p^1\ 3p^1 = \underset{\times}{\overset{\times\times}{\times P \times}}$$

and those of each isolated chlorine atom to

$$3s^2\ 3p^2\ 3p^2\ 3p^1 = \underset{\times}{\overset{\times\times}{\times Cl {\times\times}}}$$

the compound that these atoms form in consonance with the octet theory is the one with three P—Cl bonds, PCl_3.

$$\begin{array}{c} {\overset{\times\times}{Cl{\times}P{\times}Cl}} \\ {\underset{}{\overset{\times\times}{}}} \\ Cl \end{array}$$

In order to accommodate the two additional chlorine atoms in PCl_5, the phosphorous atom must clearly make available two additional half-filled orbitals, as one electron in each is required to form an individual covalent P—Cl bond. This condition may be plausibly fulfilled by assuming that one of the electrons of the unbonded $3s^2$ electron pair on the phosphorus atom is "promoted" to one of the five vacant $3d$ orbitals of the atom. This gives the orbital distribution, $3s^1\ 3p^1\ 3p^1\ 3p^1\ 3d^1$ and thus affords the necessary *five* half-filled valence orbitals for the formation of the five P—Cl bonds of the PCl_5 molecule. Manifestly, we cannot form phosphorus compounds of higher chlorination number because the phosphorus atom now provides no remaining electron pair from which a single electron can be promoted to another $3d$ orbital. Bearing in mind that the promotion of a single electron from an electron-pair makes possible two single covalent bonds (two half-filled orbitals result), the compounds, SF_2, SF_4, and SF_6 (eight, ten, and twelve electrons, respectively, on the sulfur atom) can thus be logically reconciled in terms of the number of half-filled orbitals supplied by the valence orbitals of the sulfur atom. Therefore, the following applies with respect to the involvement of the sulfur atom in the bondings, by redistribution of its electrons, as follows:

$$(S^0 = [Ne]\ 3s^2\ 3p^4)$$

valance orbitals of S in

$SF_2\left(F\overset{xx}{\underset{xx}{-}}S-F\right)$ = two, $3s^2\ 3p^2\ 3p^1\ 3p^1 \longrightarrow$ available to 2 $\left[\overset{xx}{\underset{xx}{\times}}F\times\right]$

$SF_4\left(\begin{array}{c}F\quad\quad F\\ \diagdown\overset{xx}{\diagup}\\ S\\ \diagup\quad\diagdown\\ F\quad\quad F\end{array}\right)$ = four, $3s^2\ 3p^1\ 3p^1\ 3p^1\ 3d^1 \longrightarrow$ available to 4 $\left[\overset{xx}{\underset{xx}{\times}}F\times\right]$

$SF_6\left(\begin{array}{c}F\ F\ F\\ \diagdown|\diagup\\ S\\ \diagup|\diagdown\\ F\ F\ F\end{array}\right)$ = six, $3s^1\ 3p^1\ 3p^1\ 3p^1\ 3d^1\ 3d^1 \longrightarrow$ available to 6 $\left[\overset{xx}{\underset{xx}{\times}}F\times\right]$

The premise of the expanded-octet theory upon which plausibility rests is that it is energetically feasible for an electron to be promoted to a higher energy orbital. This would seem acceptable enough were promotion from one orbital to another of only slightly greater energy — generally, of the identical principal quantum number — inasmuch as the electron must be presumed to be more stable in the lower energy orbital. Consequently, all conjectures of expansion of the octet must be considered with prudent caution.

In terms of the electron compositions of the central atom's valence orbitals, account, structurally, for
   (a) the nonexistence of SF_3 and SF_5.
   (b) the existence of NCl_3 *but not* NCl_5.
   (c) the formation of XeF_2, XeF_4, and XeF_6 on the assumption that valence-octet expansion to include the $6d$ energy sublevel is energetically feasible.

2. Arrange the following compounds in progressive sequence of decreasing polarity of the C—H bond, and explain your answer:

$$CH_2Br_2,\ CHBr_3,\ CHCl_3,\ CH_4,\ CHF_3.$$

3. We have already given attention to the relationships $\lambda\nu = c$ and $E = h\nu$, wherein emission or absorption of radiation energy is represented by a series of individually discrete and separate units, each of which is termed a *quantum*.

PROPERTIES OF CHEMICAL BOND 75

Calculate with respect to a given sample of 0.500 mole of diatomic species, AB, for which it has been determined that the rupture of a single A–B bond requires the absorption of one quantum of radiation energy:

(a) the number of quanta and the number of *einsteins* of energy needed to break all the bonds in the given sample. (1 Einstein = 1 Avogadro number of quanta.)

(b) the wavelength λ of light, in centimeters, that when absorbed provides the energy required to rupture twice as many A—B bonds as the given sample supplies. (8.03×10^4 calories of energy required for rupture of a single bond.)

(c) the frequency ν and the wave number $\bar{\nu}$ of the required light radiation.

4. It is logically to be anticipated that the shorter the internuclear distance between two oppositely charged ions in a crystal, the larger is the lattice energy of the crystal. If the radius of each component of the ion-pair, as derived from X-ray measurement of its relative position in the structure, is taken as one-half of the distance separating the two ions (a qualitatively acceptable approximation), then (in general) the sum of the two radii (positive ion and negative ion) should plausibly evaluate the relative stability of crystal structures of the same lattice type.

Consult the values of ionic radii supplied within the table of Property Periodicities of the Elements (Appendix D) and

(a) arrange the following alkali halides in the order of their diminishing crystal lattice energies:

KI, NaCl, LiBr, RbF, CsI, NaF.

(b) determine which of the alkali-halide crystal structures, plus all others not stated in the exercise (except derivatives of francium and astatine), would have, in conformity with the periodicity of size of radius, the largest lattice energy. Secondly determine which would have the smallest lattice energy.

5. The concept of *formal charge* evaluates the inequality with which electrons are distributed between a given atom and those with which it is bonded in a polyatomic ion or neutral molecule. It is intended solely as a plausible relative measurement of the excess or deficiency of electron density upon an atom that is bonded, as compared to one that is free and unbonded. In coordinate-covalent bondings, in particular, a single donor-atom that contributes both of the two electrons required for its linkage with an acceptor-atom will, in effect, have a *geographical* site of positive charge. Its acceptor-bonding partner is then oriented as a negatively charged site. The same situation holds, although in lesser degree, in normal covalent bondings, wherein the atomic partners contribute equal numbers of electrons to their mutual bonds. The electron concentrations or density shifting is in a direction toward the more electronegative partner. The method of computing formal charge, however, is distinctly different from other accounting procedures, designed to keep track of electron distributions.

Formal charge represents a way of charge evaluation which is applicable only to covalence; hence, it must not be confused with *valence* that arises from a complete transference of electrons from one atom to another in ionic bonding; nor with the *oxidation state* of an atom. Each represents a different "bookkeeping" system and, so far as it concerns oxidation state and formal charge, they are completely arbitrary. In determining an atom's oxidation state all electrons shared in a bond between the atoms are always counted fully with the more electronegative atom. In so doing, we are thus emphasizing the tendency of the covalent bond towards *ionicity*. In determining formal charge, however, the shared electrons are always divided evenly between the bonded atoms, regardless of their difference in electronegativity. Consequently, in assignments of formal charge we are stressing the tendency toward equality in the electron distribution to the respective bonding partners; that is, their tendency toward *pure covalence*. It is this differentiation of emphasis of the two types of electron-accounting procedures that reconciles the numerical variability in charge that the same atom may manifest in the same compound.

Formal charge is computed by subtracting from the total number of valence electrons held by the isolated and unbonded atom the number of electron-pair bonds and the number of unshared electrons surrounding the pertinent atom as it is oriented in the entire compound; that is

formal charge (fc) = no. valence electrons − no. unshared electrons − no. bonds.

Clearly, we might also have defined the number of bonds in the preceding formulation as one-half the total number of shared electrons.

To illustrate with the NH_4^+ ion, electronically depicted as:

$$\left[\begin{array}{c} H \\ \times\times \\ H\overset{\times}{\underset{\times\times}{\cdot}}N\overset{\times}{\cdot}H \\ H \end{array}\right]^{+}$$

In accordance with the formulation given, the following formal charges prevail therein:

$$\text{fc } N = 5 - 0 - 4 = +1$$

$$\text{fc } H = 1 - 0 - 1 = 0 \text{ (for } each \text{ hydrogen).}$$

The sum of all formal charges of all atoms in any species must be equal to the actual net charge upon that species — its valence if it is a polyatomic ion, and zero if it is a molecule. In the example given, this requirement has been satisfied; for

$$\begin{array}{r} \text{fc } N = +1 \\ 4 \times \text{fc } H = 0 \\ \hline \text{sum} = +1 \end{array}$$

which checks with the stated charge upon the ion.

Again, illustratively for the NO_3^- ion,

$$\left[\begin{array}{c} \overset{\times\times}{\underset{}{\cdot}}O\overset{\times\times}{\cdot}\alpha \\ \overset{\times\times}{\underset{\times\times}{\cdot}}O\overset{\times\times}{\cdot}N\overset{\times\times}{\cdot}O\overset{\times\times}{\cdot} \\ \alpha \beta \end{array}\right]^{-}$$

PROPERTIES OF CHEMICAL BOND   77

wherein octet requirements provide for a total of 24 electrons: 5 from the nitrogen, 6 from each of three oxygens, and the additional one acquired ionically. It is necessary to observe that different atoms of the identical element can have different formal charges in the same species, unlike oxidation state, wherein an average charge — sometimes fractional — is acquired.

Thus, with respect to the nitrate ion as electronically formulated:

$$\text{fc } O_\alpha = 6 - 6 - 1 = -1$$
$$\text{fc } O_\beta = 6 - 4 - 2 = 0$$
$$\text{fc } N = 5 - 0 - 4 = +1.$$

These formal charges add up to the specified charge (the valence) of the ion itself:

$$\begin{aligned}2 \times \text{fc } O_\alpha &= -2 \\ 1 \times \text{fc } O_\beta &= 0 \\ \underline{1 \times \text{fc } N } &= \underline{+1} \\ \text{sum} &= -1 \text{ (net)}.\end{aligned}$$

Fortified with the preceding expositions, compute the formal charge upon each of the atoms in the following species. All atoms therein are to be given completed noble gas configurations that conform to the indicated single, double, and triple bonds.

(a) $[H_2N \overset{\times}{\underset{\times}{}} NH_2]^0$.

(b) $[H_3N \overset{\times}{\underset{\times}{}} BF_3]^0$.

(c) $[\overset{\times\times}{\underset{\times\times}{N}} \overset{\times\times}{\underset{\times\times}{C}} \overset{\times\times}{\underset{\times\times}{O}}]^-$.

(d) $[H \overset{\times\times\times}{\underset{\times\times\times}{C}} \overset{\times\times\times}{\underset{\times\times\times}{N}}]^0$.

(e) $\left[\overset{\times\times\times\times\times\times\times\times}{\underset{\times\times\times\times\times\times\times\times}{\text{Cl}\overset{\times}{\underset{\times}{S}}\overset{\times}{\underset{\times}{S}}\text{Cl}}} \right]^0$.

(f) $\left[H \overset{\times\times}{\underset{}{N}} \overset{}{\underset{}{N}} \overset{\times\times}{\underset{}{N}} \atop \alpha \beta \gamma \right]^0$.

(g) $\left[\begin{array}{c} \overset{\times\times}{H \overset{\times}{\underset{\times}{O}} H} \\ H_3C \overset{\times\times\times\times}{\underset{\times\times}{C}} C \overset{\times\times}{\underset{}{O}} \\ \overset{\times}{\underset{\times}{N}}H_2 \end{array} \right]^0$.

(h) $[BF_4]^-$.

6. Ozone, O_3, is a diamagnetic molecule (no unpaired electrons). Its bent structure ($\measuredangle \approx 127°$) has been ascertained to encompass two bonds of identical length — each equivalent to 1.5 bonds.

Provide the electron-dot structures of the contributory resonance hybrids.

7. The polyatomic ion, I_3^-, is formed by reaction between molecular iodine and an alkali halide. In the linear structure thereof, the formal charge of the central iodine atom is of -1.

(a) Supply the electron-dot structure of this ion, showing unambiguously the correct numbers of outer-shell electrons associated with the over-all structure.

(b) Give the respective numbers of unshared electrons, of unshared electron-pairs, and of electron-pair bonds associated with the valence shell of the central atom.

(c) What is the formal charge on either one of the outer iodine atoms?

CHAPTER
THREE

COMPLEXING TRANSITIONAL METALS

THE PRINCIPAL CHARACTERISTICS OF CHEMICAL BEHAVIOR THAT CONSTITUTE the prime objectives of this chapter are those that involve the formation of metallic complex ions and, in some instances, of neutral uncharged complexes. The required considerations are very largely those of transition-metal ions which utilize *d* orbitals for bonding purposes. The first task in interpreting them, therefore, is to recall and to elaborate further the already-established fundamental concepts of quantum mechanics as they concern the availability of such orbitals and of the electrons that they potentially are able to accommodate.

ORBITAL COMPOSITIONS OF SOME TRANSITIONAL AND POST-TRANSITIONAL ELEMENTS

Our illustrations here of the rules already presented are confined to un-excited atoms and simple free cations of the elements scandium (atomic number 21) to zinc (atomic number 30) inclusive. Each member in this sequential series of progressive build-up of orbital electrons has the *argon* core — $1s^2$, $2s^2$, $2p^6$, $3s^2$, $3p^6$ — common to all members of the fourth row in the periodic classification. The superscripts to the sublevels denote, conveniently, the total number of electrons in all orbitals thereof. As all d orbitals are energetically equivalent in the free atom or ion that is unperturbed by the presence or approach of ligands, no differentiations are needed with respect to the orbital designations, d_{xz}, d_{xy}, d_{yz}, $d_{x^2-y^2}$, or d_{z^2}.

We pay heed in development here to the numbers of unpaired electrons made available. In validating the concept of maximum utilization of orbitals, which requires that electrons remain unpaired in parallel-spin as long as possible before they pair in opposed-spin (Hund's rule of maximum multiplicity), an awareness of the numbers of unpaired electrons assists most helpfully in predicting the spatial geometries or *stereochemistry* of complexes. This represents a prime, ultimate consideration of the development undertaken so far.

Measurements of the *magnetic moment* of a species are calculated from data obtained with the Gouy balance. This device evaluates the magnitude of the attraction to its electromagnetic field of paramagnetic substances, and of the repulsion of diamagnetic substances. The simple calculation of magnetic moment (μ_M) in units of *Bohr magnetons* conforms to the data shown in Table 3.1,

$$\mu_M = \sqrt{n^2 + 2n} \text{ Bohr magnetons}$$

where n represents the total number of unpaired electrons in all of the separate and distinct orbitals.

In the table, Z is the symbol for *atomic number*, and the data in parentheses following the chemical symbols denote the numbers of unpaired electrons (as experimentally verified) and the magnetic moment μ_M, respectively. It may be observed, as well, that the $4s$ electrons are always lost in ion-formation before those of $3d$.

Note the adjustments undergone in $4s$ and $3d$ orbital composition of the neutral chromium atom — in consonance with the experimentally reconciled greater stability of the orbitals on each sublevel when they are all exactly half-filled. Were the neutral chromium atom to have the sequential composition, $3d^1$ $3d^1$ $3d^1$ $3d^1$ $3d^0$ $4s^2$, the paramagnetic equivalent would correspond to four unpaired electrons and a magnetic moment of $\sqrt{(4)^2 + (2 \times 4)}$, or 4.90 Bohr units. This would contradict the measured values of five unpaired electrons, and of a magnetic moment of $\sqrt{(5)^2 + (2 \times 5)}$ or 5.92 Bohr units.

Note, likewise, the adjustments undergone in the orbital composition of the neutral copper atom. It conforms to the greater stability of a fully completed set of $3d$ orbitals and a half-filled $4s$ orbital. The single unpaired electron is found, consequently, in the $4s$ sublevel rather than in the $3d$ sublevel, thus yielding an electron configuration of $3d^{10}$ $4s^1$ instead of the otherwise expected configuration of $3d^9$ $4s^2$.

Table 3.1

Magnetic moment calculated in Bohr magnetons

Element	3d	3d	3d	3d	3d	4s	Ion (unpaired; μ_M)
scandium, z 21: [Ar core]	$3d^1$	$3d^0$	$3d^0$	$3d^0$	$3d^0$	$4s^2$	$\rightarrow Sc^0(1; \mu_M = 1.73)$
	$3d^0$	$3d^0$	$3d^0$	$3d^0$	$3d^0$	$4s^0$	$\rightarrow Sc^{III}(0; \mu_M = 0.00)$
titanium, z 22: [Ar core]	$3d^1$	$3d^1$	$3d^0$	$3d^0$	$3d^0$	$4s^2$	$\rightarrow Ti^0(2; \mu_M = 2.83)$
	$3d^1$	$3d^1$	$3d^0$	$3d^0$	$3d^0$	$4s^0$	$\rightarrow Ti^{II}(2; \mu_M = 2.83)$
	$3d^1$	$3d^0$	$3d^0$	$3d^0$	$3d^0$	$4s^0$	$\rightarrow Ti^{III}(1; \mu_M = 1.73)$
	$3d^0$	$3d^0$	$3d^0$	$3d^0$	$3d^0$	$4s^0$	$\rightarrow Ti^{IV}(0; \mu_M = 0.00)$
vanadium, z 23: [Ar core]	$3d^1$	$3d^1$	$3d^1$	$3d^0$	$3d^0$	$4s^2$	$\rightarrow V^0(3; \mu_M = 3.87)$
	$3d^1$	$3d^1$	$3d^1$	$3d^0$	$3d^0$	$4s^0$	$\rightarrow V^{II}(3; \mu_M = 3.87)$
	$3d^1$	$3d^1$	$3d^0$	$3d^0$	$3d^0$	$4s^0$	$\rightarrow V^{III}(2; \mu_M = 2.83)$
	$3d^1$	$3d^0$	$3d^0$	$3d^0$	$3d^0$	$4s^0$	$\rightarrow V^{IV}(1; \mu_M = 1.73)$
	$3d^0$	$3d^0$	$3d^0$	$3d^0$	$3d^0$	$4s^0$	$\rightarrow V^V(0; \mu_M = 0.00)$
chromium, z 24: [Ar core]	$3d^1$	$3d^1$	$3d^1$	$3d^1$	$3d^1$	$4s^1$	$\rightarrow Cr^0(6; \mu_M = 6.93)$
	$3d^1$	$3d^1$	$3d^1$	$3d^1$	$3d^0$	$4s^0$	$\rightarrow Cr^{II}(4; \mu_M = 4.90)$
	$3d^1$	$3d^1$	$3d^1$	$3d^0$	$3d^0$	$4s^0$	$\rightarrow Cr^{III}(3; \mu_M = 3.87)$
	$3d^1$	$3d^1$	$3d^0$	$3d^0$	$3d^0$	$4s^0$	$\rightarrow Cr^{IV}(2; \mu_M = 2.83)$
	$3d^0$	$3d^0$	$3d^0$	$3d^0$	$3d^0$	$4s^0$	$\rightarrow Cr^{VI}(0; \mu_M = 0.00)$
manganese, z 25: [Ar core]	$3d^1$	$3d^1$	$3d^1$	$3d^1$	$3d^1$	$4s^2$	$\rightarrow Mn^0(5; \mu_M = 5.92)$
	$3d^1$	$3d^1$	$3d^1$	$3d^1$	$3d^1$	$4s^0$	$\rightarrow Mn^{II}(5; \mu_M = 5.92)$
	$3d^1$	$3d^1$	$3d^1$	$3d^1$	$3d^0$	$4s^0$	$\rightarrow Mn^{III}(4; \mu_M = 4.90)$
	$3d^1$	$3d^1$	$3d^1$	$3d^0$	$3d^0$	$4s^0$	$\rightarrow Mn^{IV}(3; \mu_M = 3.87)$
	$3d^1$	$3d^0$	$3d^0$	$3d^0$	$3d^0$	$4s^0$	$\rightarrow Mn^{VI}(1; \mu_M = 1.73)$
	$3d^0$	$3d^0$	$3d^0$	$3d^0$	$3d^0$	$4s^0$	$\rightarrow Mn^{VII}(0; \mu_M = 0.00)$
iron, z 26: [Ar core]	$3d^2$	$3d^1$	$3d^1$	$3d^1$	$3d^1$	$4s^2$	$\rightarrow Fe^0(4; \mu_M = 4.90)$
	$3d^2$	$3d^1$	$3d^1$	$3d^1$	$3d^1$	$4s^0$	$\rightarrow Fe^{II}(4; \mu_M = 4.90)$
	$3d^1$	$3d^1$	$3d^1$	$3d^1$	$3d^1$	$4s^0$	$\rightarrow Fe^{III}(5; \mu_M = 5.92)$
	$3d^1$	$3d^1$	$3d^0$	$3d^0$	$3d^0$	$4s^0$	$\rightarrow Fe^{VI}(2; \mu_M = 2.83)$
cobalt, z 27: [Ar core]	$3d^2$	$3d^2$	$3d^1$	$3d^1$	$3d^1$	$4s^2$	$\rightarrow Co^0(3; \mu_M = 3.87)$
	$3d^2$	$3d^2$	$3d^1$	$3d^1$	$3d^1$	$4s^0$	$\rightarrow Co^{II}(3; \mu_M = 3.87)$
	$3d^2$	$3d^1$	$3d^1$	$3d^1$	$3d^1$	$4s^0$	$\rightarrow Co^{III}(4; \mu_M = 4.90)$
	$3d^1$	$3d^1$	$3d^1$	$3d^1$	$3d^1$	$4s^0$	$\rightarrow Co^{IV}(5; \mu_M = 5.92)$
nickel, z 28: [Ar core]	$3d^2$	$3d^2$	$3d^2$	$3d^1$	$3d^1$	$4s^2$	$\rightarrow Ni^0(2; \mu_M = 2.83)$
	$3d^2$	$3d^2$	$3d^2$	$3d^1$	$3d^1$	$4s^0$	$\rightarrow Ni^{II}(2; \mu_M = 2.83)$
	$3d^2$	$3d^2$	$3d^1$	$3d^1$	$3d^1$	$4s^0$	$\rightarrow Ni^{III}(3; \mu_M = 3.87)$
	$3d^2$	$3d^1$	$3d^1$	$3d^1$	$3d^1$	$4s^0$	$\rightarrow Ni^{IV}(4; \mu_M = 4.90)$
copper, z 29: [Ar core]	$3d^2$	$3d^2$	$3d^2$	$3d^2$	$3d^2$	$4s^1$	$\rightarrow Cu^0(1; \mu_M = 1.73)$
	$3d^2$	$3d^2$	$3d^2$	$3d^2$	$3d^2$	$4s^0$	$\rightarrow Cu^I(0; \mu_M = 0.00)$
	$3d^2$	$3d^2$	$3d^2$	$3d^2$	$3d^1$	$4s^0$	$\rightarrow Cu^{II}(1; \mu_M = 1.73)$
zinc, z 30 [Ar core]	$3d^2$	$3d^2$	$3d^2$	$3d^2$	$3d^2$	$4s^2$	$\rightarrow Zn^0(0; \mu_M = 0.00)$
	$3d^2$	$3d^2$	$3d^2$	$3d^2$	$3d^2$	$4s^0$	$\rightarrow Zn^{II}(0; \mu_M = 0.00)$

DEFINITIONS OF TERMS

Some clarification is, perhaps, due with respect to the distinction not always observed, between the terms *transitional* and a *post-transitional* species. Both comprise fully the elements that lie between the main group classifications II and III of the long form of the Periodic Table. As such, they collectively and restrictively represent the fourth-row, argon-core, elements of Z 21 to Z 30 inclusive (ten elements), the fifth-row, krypton-core, elements of Z 39 to Z 48 inclusive (ten elements), and certain of the sixth-row, xenon-core, elements — actually, again ten in number — lanthanum of Z 57 and (after an interruption for the rare earth elements or *lanthanides*), the additional nine of Z 72 to Z 80 inclusive.

Transitional species are characterized by the presence of partly vacant d orbitals. Their electron configurations thus invite a belated filling of yet-incomplete d orbitals by attractions of ligands, with the consequent formation of complexes (ionic or neutral).

Post-transitional species have their d orbitals already fully occupied but, by circumstance of construction of the Periodic Table, are likewise found between, and are, hence, excluded from the main periodic classification groups.

Properly to be classified in the post-transitional category are elemental copper (Cu^0; $3d$ orbitals, complete), silver (Ag^0; $4d$ orbitals, complete), and gold, (Au^0; $5d$ orbitals, complete). Similarly post-transitional are elemental zinc (Zn^0; $3d$ orbitals, complete), cadmium (Cd^0; $4d$ orbitals, complete), and mercury (Hg^0; $5d$ orbitals, complete). The terminological distinctions that are made, however, become quite tenuous when it is observed that a post-transitional free metal can yield a cation of transitional behavior. Thus, examination of the orbital electron configurations of the copper species that were provided earlier shows both elemental Cu^0 and its monovalent cation Cu(I) to be post-transitional forms, whereas the divalent Cu(II) is a transitional cation. The vagaries of definition here should hardly be permitted to become critical, particularly because many other elements that have fully completed d orbitals, and which lie between the two main periodic groups II and III are not to be classified as post-transitional. These include the *lanthanides* (or rare earth elements), characterized by the presence of partially filled $4f$ orbitals — and the *actinides* — characterized by the presence of partially filled $5f$ orbitals. Neither transitional nor post-transitional metals are involved in f-orbital filling.

INTERPRETATIONS OF BONDING IN TRANSITION COMPLEXES

Applied or prevaling theories of chemical bonding in transition complexes may be represented as four in number:

1. Molecular Orbital theory (MO) (J. van Vleck, 1935). Essentially, the overlap and ensuing coalescence of hitherto separate atomic orbitals under stabilization "inducements."

This results in the formation of a new orbital — the *molecular orbital* — with the electron pair being influenced by two nuclei. Since the molecular orbital becomes, thus, a multiple-nuclei orbital, the relative densities of electron distribution and, consequently, the shape of the orbitals, differ from those of the atomic orbitals before they coalesced. (These considerations have already been explained in Chapter 2.)

2. Electrostatic Crystal Field Theory (CF) (H. Bethe, 1929). A representation of bonding that is purely an electrostatic or coulombic interaction between positively charged and negatively charged species. Complexes are thus presumed to form when centrally situated cations electrically attract *ligands*, that is, anions or dipole molecules. This is because the cations are positively charged whereas anions are negatively charged; and the dipole molecules, as well, can offer their negatively incremented ends for such electrostatic attractions.

3. Ligand Field Theory (LF). Essentially the electrostatic crystal field theory with added, but relatively minor, concessions made to molecular orbital concepts of overlap.

4. Valence Bond Theory (VB) (L. Pauling, J. L. Slater, 1935). Essentially, the concept of amalgamation of atomic orbitals of differing energy states into energetically equivalent "hybrid" orbitals. Hybridization of orbitals to form complex ions or neutral complexes may validly be viewed as the peripheral overlap of filled orbitals of ligands with vacant orbitals of a centrally situated metallic atom or cation.

In keeping with the mechanistic theme of this chapter which delineates the constitutional nature of species, the necessarily limited objectives here are the explanatory applications, to complexing, of the qualitatively satisfying electrostatic crystal field theory — as a preferential alternative to classic valence bond concepts.

Because the electrostatic or coulombic attractions of the crystal field theory concern the availability of unfilled d orbitals, we necessarily confine ourselves to the metals or ions of transitional character. Complexing of species of other types which preclude d-orbital filling but which, nonetheless, exhibit significant — at times, even profound — capacities for complexing, may not be ignored. Such species, however, for example the post-transitional Zn^{2+} ion — may be amply interpreted with respect to its complexing propensities by the molecular orbital and/or valence bond theories.

The plethora of bonding concepts makes the subject of complexing, as well as the selection of the particular theory itself, somewhat controversial. Because, moreover, any ambitions to undertake an intensive delineation of all the theories are automatically precluded by the limited objectives set for this text, some explanation is due for the preference given here to the electrostatic crystal field theory as a means for comprehending the qualitative mechanisms of complexing of transition-metal ions.

For one thing, the classic valence bond approach is given adequate attention in all modern general chemistry textbooks as the plausible and useful technique of normal and coordinate covalent attachments of species. Some of these have already been presented in Chapter 2. However, the projection of the valence bond (VB) theory to transition-metal ion complexing has revealed certain inadequacies in accounting quantitatively for measured spectral and magnetic properties of complexed species. The continual "patching" that the VB theory has undergone in attempts to maintain its utility against obvious quantitative inconsistencies in this type of bonding has impaired its directness as a qualitative tool, as well. As an alternative, the CF theory, although it, itself, is purely qualitative, injects no extraneous issues that must subsequently be compromised. Actually, the CF theory, when augmented with partial considerations of the overlapping of atomic orbitals to form molecular orbitals, serves *quantitative* objectives as well, when the effects of such orbital overlap upon the complex as a whole are given careful mathematical weight. This combination of the CF theory with partial concessions made to molecular orbital concept constitutes what generally is (or was, originally) called ligand field (LF) theory. The distinction between the two has proved to be not too critical, however, inasmuch as the two terms of reference are now quite loosely interchanged.

The molecular orbital theory is by far the most complete, the most exact, and the most widely applicable treatment of bonding. Quantitative measurements of spectral and magnetic properties of species check out most successfully when viewed in terms of MO bonding. The mathematical involvements and inconveniences of MO theory when applied to complexes in general are, however, ordinarily profound enough to discourage its applicability thereto in favor of the far simpler CF, or VB theory. The molecular orbital theory remains generally reserved for its more convenient qualitative expositions of bonding in diatomic molecules and in organic aromatic hydrocarbons, as well as in all instances where quantitatively exact measurements of magnetic and spectral properties of species are desired.

The empirical representations of the crystal field theory in explaining bonding in transition complexes — that is, the purely electrostatic mutual attractions between oppositely-charged species — should not beguile one away from a realization that a significant degree of bonding covalence (normal or coordinate) may, nonetheless, exist. The overlapping of atomic orbitals to form a new molecular orbital, which basically underlies the valence bond theory, as well as the molecular orbital theory, represents the modifying concession to exactitude that is offered by the otherwise electrostatic ligand field theory.

COORDINATION NUMBER; LIGANDS

As it relates to transition and post-transition complexes, the "coordination sphere" represents the collective entity or unit made up of the centrally oriented, positively charged metal ion to which are coordinated (not necessarily coordinate covalent bonding) groups of either negatively charged ions (anions) or dipole molecules. The latter are coordinated by virtue of steric factors — the sites upon the molecules that bear increments of negative charge. Coordinated species of whatever nature are known as *ligands*; hence, the terms "coordination" and "ligancy" are synonymous. The number of anions and/or dipole molecules that may be coordinated to the central metal (that is, the *coordination number* or *ligancy*) is determined by

(1) *the nature of the ligands.* The accommodating of ligands within the coordination sphere, clearly, is greatly influenced by the sizes of the ligands and by the repulsions that exist between and among their electron clouds. Since electrostatic attractions may invite ligands of different sizes and degrees of repulsion, to the same sphere, there may be variability in the coordination number of ions of identical transition metals. Such variability in the coordination number of the same central metal ion in different complexes, however, must not be considered to refute the *constancy* of its coordination number prevailing with a specific unvarying set of ligands characterizing a particular complex.

(2) *the nature of the centrally-oriented metal-ion.* The size and cationic charge of a centrally oriented metallic ion must inevitably greatly affect the number of ligands that it may draw in to its sphere of coordination. In general, the greater the cationic charge, the larger is the number of ligands permitted, assuming that such comparisons are made with chemically identical ligands. Likewise, quite logically, the larger the size of the cation, the greater is the coordination number of the cation — once again, with comparisons delineated for similar ligands.

It is necessary to stress at this point that the coordination number is *not* to be defined or interpreted as the total number of ligands. Rather, it is the total number of atoms (of the ligands) provided with available coordination sites on the central positive element and, therefore, numerically representative of the total of chemical bonds formed. It is important to appreciate this, inasmuch as a single ligand may occupy more than one site upon the central atom. Thus, although frequently the ligancy does correspond numerically to the total number of ligand molecules and/or anions present in the sphere of coordination, in some instances the ligancy or coordination number actually exceeds the number of ligands. When more than one atom of a single ligand molecule or ion occupies an available coordination position it may be presumed to bend itself pincer-like around the central atom to form a complex ring-structure called a *chelate* (from Greek for *claw*). Polydentate (literally, *many teeth*) ligands are referred to as *bidentate, triden-*

tate, tetradentate, pentadentate, or *hexadentate,* according to their respective attachments in two, three, four, five, or six positions in the sphere of coordination.

RADIUS RATIOS

The numerical evaluations of the relative sizes of coordinating cation and coordinated ligand prove useful in determining, fairly accurately, the coordination number. Moreover, as the coordination number represents the number of pairs of sigma (σ)-bonds that are primarily responsible for the shape of a complex, we are most opportunely provided with the means to predict therefrom the particular shape of a species. In using the *radius ratio rule,*

$$\text{radius ratio} = \frac{\text{Radius of cation}}{\text{Radius of ligand}}$$

we make the assumption that ligands of similar size are packed tightly around, and are in direct contact with, a centrally situated spherical, rigid cation. The concept that there is no distortion in an atom that is subjected to attractive and/or repulsive forces of the electric charges of its neighboring atoms is a quite idealized assumption. It will, however, prove adequately serviceable for the useful approximations that may be made with respect to the maximum limits of ligancy and the shape of the structure to which the pertinent coordination number predisposes the complexed species. These are defined in Table 3.2.

Table 3.2

Relationships of ligands

Radius Ratio	Coordination Number	Shape of Complex
>0.000–0.155	2	linear
>0.155–0.225	3	triangular planar
>0.255–0.414	4	tetrahedral
>0.414–0.590	4	square planar
>0.414–0.732	6	octahedral
>0.645	8	antiprismatical
>0.732	8	cubic

Species of other coordination or ligancy, and of other shapes, are also known. The most frequently encountered, however, are those of 2-coordination (linear), 4-coordination (both square planar and tetrahedral), and 6-coordination (octahedral).

EXPERIMENTAL BASIS FOR COORDINATION-SPHERE CONCEPTS

Literally, thousands of coordination compounds are known. The experimental observations and theoretical postulates of *coordination chemistry* of A. Werner (Switzerland, 1893) provided the impetus to recognizing them, and to continued intensive investigation, particularly of the chemical bonding forces involved.

As plausibly conceived, a cation has two kinds of valence simultaneously; a *primary* or *electrovalence*, and a *secondary* or *auxiliary valence* by which the cation attracts to itself additional species (ligands) despite the fact that its primary valence is already fully satisfied. It should be apparent from what has preceded in our present development, as well as what follows, that the secondary or auxiliary valence is denoted by the number of chemical bonds that the central metal ion forms with anions and/or molecules within its sphere of coordination. The number of these bonds equals, as already observed, the coordination number or *ligancy* of the metal in question.

Werner's laboratory studies with solutions of three different ammonia derivatives of cobaltic chloride — compositions formulated as $CoCl_3 \cdot 6NH_3$, $CoCl_3 \cdot 5NH_3$, and $CoCl_3 \cdot 4NH_3$, respectively — elicited the following observations and conclusions:

(1) Neither H_2SO_4 nor NaOH *readily* interacted with the NH_3 content of the solutions. As a strong acid normally easily neutralizes a weak base, and a strong base normally easily displaces a weak base, apparently the NH_3 molecules are rather tightly bound (coordinated).

(2) Electrolysis of each of the solutions resulted in the discharge at the anode (the positive electrode in the electrolytic cell) of chlorine gas, Cl_2. At the cathode, however, (the negative electrode in the electrolytic cell) not only was cobalt metal discharged, but also ammonia gas, NH_3. As positively charged ions (cations) migrate to the negative electrode, the electropositive cobalt and the electroneutral ammonia appear to be associated together in a positively charged aggregate, or complex ion.

(3) The addition of excess silver nitrate ($AgNO_3$) to separate equimolar solutions of the three salts yielded different quantities of precipitated silver chloride, AgCl. This occurred despite the identical number of chlorine atoms in the three separate solutions of the salts, as formulated. Clearly, the Cl^- ion is available in different concentrations for such precipitation, notwithstanding the numerical equality of gram-formula content of chlorine in a mole of each salt.

(4) Measurements of conductivity revealed different molar conveyances of the electric current. As it is the ions that carry the current, each solution provides different numbers of ions in a mole of the parent solute.

(5) *Colligative* properties of the respective solutions (freezing point, boiling point, vapor pressure, osmotic pressure), which, like electrical conductivity, reflect the numbers of ions present, regardless of chemical identity, yielded different results for each solution, although all were of identical formal concentration. Again, indisputably, the numbers of available ions must differ for the different solutions.

Some of the experimental data mentioned in the foregoing are given in Table 3.3. These afford logical reconciliation of the formulations that subsequently are made of the respective dissolved compounds in terms of their coordination spheres; that is the manifested secondary valence of the centrally-situated electropositive metallic cobalt. It also gives the ionic composition and charge external to the coordination spheres; these external arrangements manifest the principal valence of the centrally-situated electropositive metallic cobalt. To enhance the total picture somewhat, we also include a fourth ammonia derivative of the cobaltic chloride that was investigated subsequent to Werner's initial work — namely, the compound, $CoCl_3 \cdot NH_3$.

In all these compounds the principal valence of the cobalt remains numerically three. Its algebraic contribution to the over-all net charge upon each of the complexes that it forms is precisely $+3$ because the electroneutral molecules of NH_3 in the respective compounds offer zero charge and each Cl^- ion offers a -1 charge. The secondary valence of the cobalt remains numerically six in all of the compounds, inasmuch as in each instance it has formed six bonds with the total of six ligands represented within the coordination sphere — as shown by the bracketed formulas.

In conformity with the experimental evaluations shown in Table 3.3 we derive the following:

Formula

compound a: $[Co(NH_3)_6]^{3+}3Cl^-$ or $[Co(NH_3)_6]Cl_3$
compound b: $[Co(NH_3)_5Cl]^{2+}2Cl^-$ or $[Co(NH_3)_5Cl]Cl_2$
compound c: $[Co(NH_3)_4Cl_2]^+Cl^-$ or $[Co(NH_3)_4Cl_2]Cl$
compound d: $[Co(NH_3)_3Cl_2]^0$, a non-ionic neutral complex

Table 3.3

	Compound			
Interpretative Reference	a $CoCl_3 \cdot 6NH_3$	b $CoCl_3 \cdot 5NH_3$	c $CoCl_3 \cdot 4NH_3$	d $CoCl_3 \cdot 3NH_3$
fraction of Cl precipitated by Ag^+ ion, per mole of compound	3/3	2/3	1/3	0/3
moles of free Cl^- ion available, per mole of compound	3	2	1	0
moles of all ions available, per mole of compound—in accordance with electrical conductivity and freezing point	4	3	2	0

COMPLEXING TRANSITIONAL METALS

Table 3.4 shows a more extensive series of coordination complexes involving platinum(IV), similarly appraised in terms of the conformity of assigned coordinations with the stated experimental observations.

The identities of the ions and compounded formulations that correspond to the experimental evaluations shown in Table 3.4 are as follows:

Formula

compound a: $[Pt(NH_3)_6]^{4+}4Cl^-$ *or* $[Pt(NH_3)_6]Cl_4$
compound b: $[Pt(NH_3)_5Cl]^{3+}3Cl^-$ *or* $[Pt(NH_3)_5Cl]Cl_3$
compound c: $[Pt(NH_3)_4Cl_2]^{2+}2Cl^-$ *or* $[Pt(NH_3)_4Cl_2]Cl_2$
compound d: $[Pt(NH_3)_3Cl_3]^+Cl^-$ *or* $[Pt(NH_3)_3Cl_3]Cl$
compound e: $[Pt(NH_3)_2Cl_4]^0$, a non-ionic neutral complex
compound f: $K^+[Pt(NH_3)Cl_5]^-$ *or* $K[Pt(NH_3)Cl_5]$
compound g: $2K^+[PtCl_6]^{2-}$ *or* $K_2[PtCl_6]$

Table 3.4

Interpretative Reference	Compound						
	a $PtCl_4 \cdot 6NH_3$	b $PtCl_4 \cdot 5NH_3$	c $PtCl_4 \cdot 4NH_3$	d $PtCl_4 \cdot 3NH_3$	e $PtCl_4 \cdot 2NH_3$	f $KPtCl_5 \cdot NH_3$	g K_2PtCl_6
number of free Cl^- ions per formula unit of compound	4	3	2	1	0	0	0
number of all ions per formula unit of compound	5	4	3	2	0	2	3

The principal valence of the platinum remains numerically four in all these compounds; the secondary valence remains likewise constant, at a coordination number of six.

The value of the Werner coordination theory is, perhaps, most amply demonstrated in the highly successful, experimentally validated predictions it has afforded of the many existent geometric isomers of complexes.

GEOMETRIC SYMMETRY: STEREOISOMERISM OF THE COORDINATION SPHERE

The more common spatial orientations of complexes are illustrated in Figure 3.1 by means of chemically representative species.

The two forms for the palladium complex [Fig. 3.1 (*c*)] arise from the difference in the identities of the coordinating groups. Clearly, the distances between the two Cl, or between the two NH_3 groups, in the *trans-* form are different from those in the *cis-* form. In the *trans* form, so called because the two identical factors, either Cl or NH_3, are at diagonally opposite corners

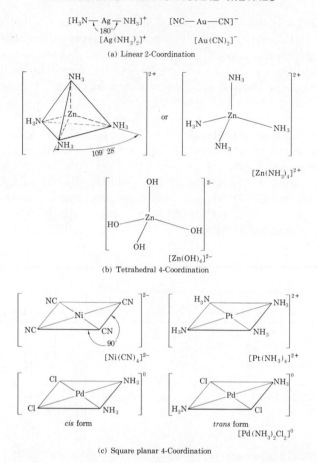

FIG. 3.1 Spatial orientations of complexes. (a) Linear 2-coordination. (b) Tetrahedral 4-coordination. (c) Square planar 4-coordination.

of the square and the distance between similar ligands corresponds to the hypotenuse of a right triangle. In the *cis*- form, the two similar ligands are at corners on the same side of the square. If the distance between the two identical ligands is denoted by d, the distance between the same two ligands in the *trans*- form is the square root of 2d² (i.e., $\sqrt{2d^2}$), because, mathematically, the square of the hypotenuse is equal to the sum of the squares of both arms of a right triangle.

Although *mixed* ligands in square planar 4-coordination are thus not equidistant from one another, they, nevertheless, all remain virtually equidistant from the centrally placed metal. Any qualification in such interpretation that may arise with mixed ligands obviously is a concession to the relatively slight distortions of symmetry that occur when radius ratios and ligand repulsions are not all identical.

The type of isomerism just described is known as *geometric* or *"cis-trans" isomerism*. As such, geometric isomers thus have identical formulas and identical types of chemical bonds. The differences in the spatial arrangements of their atoms, however, cause them to be distinctly different compounds with different sets of physical and chemical properties.

It is to be noted that geometric isomerism is not possible with tetrahedral 4-coordination complexes and certainly not with linear 2-coordination symmetry. Regardless of the number of mixed ligands in a tetrahedral structure, the distances between any or all of them remain virtually the same, although, again, some distortions of symmetry inevitably occur when the ligands are not all identical. These relatively slight distortions have already been cited as revealing differences in radius ratios and in the extents of ligand repulsions. The student can resolve this challenge to three-dimensional visualization of the absence of geometric isomerism in tetrahedral 4-coordination by pondering the symmetry of the tetrahedral structure, See Figure 3.1 (*b*).

Another form of stereoisomerism also observed is *optical isomerism*. This type of spatial orientation involves the mirror image ("right hand/left hand") differences of species and their opposing rotations of the plane of polarized light that is, light from vibrations entirely in a single plane. Optical isomerism, however, does not concern us here.

OCTAHEDRAL 6-COORDINATION

Complexes that are 6-coordinated are always octahedral in shape; and geometric isomerism exists very definitely in the 6-coordination symmetry of the three-dimensional *octahedral* structure. As already noted, some distortions of symmetry arise when all the ligands are not identical chemically, and resultant radius ratios and ligand repulsions are, consequently, not all precisely similar. Complexes in octahedral symmetry are shown in Figure 3.2 (*a*).

As with the symmetry in square planar 4-coordination, geometric isomers are also possible with octahedral 6-coordination complexes, when ligands are of mixed variety. Thus, for the complex cation $[Co(NH_3)_4Cl_2]^+$, two isomeric forms exist — a green-colored *trans*-isomer, and a violet-colored *cis*-isomer. We picture these as shown in Figure 3.2 (*b*).

Here, again, the student must exercise imagination in translating into proper three-dimensional perspective what can be superficially pictured in only two dimensions on paper. Were we to number the positions of the ligands of the structure as shown in Figure 3.2 (c), it should be recognized that the pairs 1–2, 1–3, 1–4, 1–5, 2–3, 2–5, 2–6, 3–4, 3–6, 4–5, 4–6, and 5–6

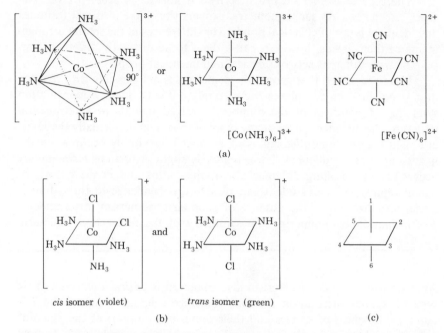

FIG. 3.2   Octahedral 6-coordinated complexes. $[Co(NH_3)_6]^{3+}$; $[Fe(CN)_6]^{4-}$ symmetry. (b) *Cis-trans* isomerism. (c) Perspective in positions of ligands.

are all equivalent *cis* positions. On the other hand, the pairs 1–6, 2–4, and 3–5 are all equivalent *trans* positions.

In delineating stereoisomerism as an attribute solely of the coordination sphere, we must avoid any inadvertent neglect of the species that reside outside the coordination sphere and which, in conjunction with it, constitute the entire compound.

A few additional examples of isomerism may well be taken in terms of reference to a specific compound as a whole; that of $Cr(H_2O)_6Cl_3$ for instance. Conforming to this identical formula are three different known compounds. One, violet-colored, corresponds to the formulation $[Cr(H_2O)_6]Cl_3$. The other two, both green, correspond to the separate formulations $[Cr(H_2O)_5Cl]Cl_2 \cdot H_2O$ and $[Cr(H_2O)_4Cl_2]Cl \cdot 2H_2O$. Clearly, the interpretation of their varying physical and chemical properties is to be found in the construction of their respective coordination spheres, as well as, concomitantly, in the pertinent species of the compound external to the coordination sphere. The perplexities of terminological reference in such cases can be resolved only by the convenience of a systematized nomenclature. This is provided shortly.

Modern quantitative theoretics that reconcile the stability of complexes with their spatial arrangements suggest that most, if not all, the square

COMPLEXING TRANSITIONAL METALS

planar 4-coordinations are really three-dimensional octahedral 6-coordinations. As such, by elongating one of the Cartesian axes it may be imagined that attachment-site "privileges" are afforded to *two* loosely-held molecules of water — assuming the complex to be either a hydrate of solid form or dissolved in water. The latter represents our major interest. Consequently, we may plausibly relate an observed square planar orientation of the complex $[ML_4]^{2-}$ as formed between the generalized central cation M^{2+} and the ligand anions, L^-, to a theorized actuality of octahedral 6-coordination as shown in Figure 3.3.

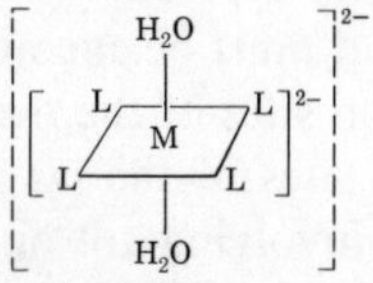

FIG. 3.3 Square planar and octahedral reconciliations of an aquo complex.

GEOMETRY IN CHELATES

Chelates and *chelation* have been defined earlier. Fortified with the introduction to the general geometry of complexes already presented, the student may logically interpret the following chelates and chelating agents in their mutual frames of spatial orientations. The few illustrations here (Fig. 3.4) are confined to bidentate chelating ligands, specifically, the ethylenediamine molecule, $NH_2CH_2CH_2NH_2$ (abbreviated *en*), and the oxalate ion, $C_2O_4^{2-}$,

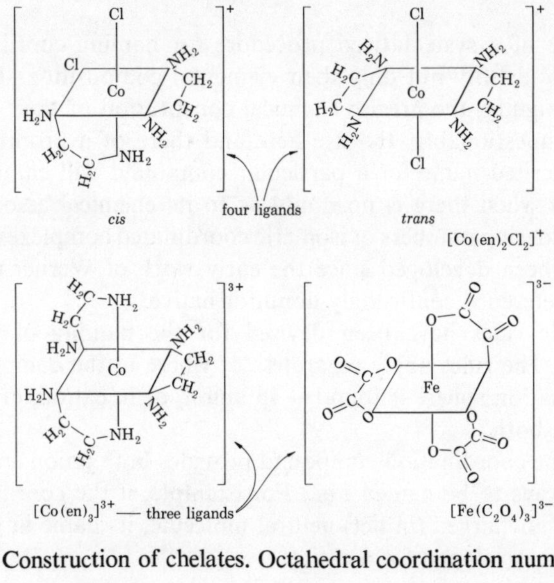

FIG. 3.4 Construction of chelates. Octahedral coordination number of 6.

in their separate employments in octahedral-symmetry complexing. The cations in the illustrations are represented by Co(III) and Fe(III). Bidentate chelating agents offer *two* sites for attachment to the single centrally located metallic ion, with consequent formation of ring structures.

A cautious reserve must be exercised in predicting the reaction exchanges of chelating ligands and of the existence of *trans* forms of the resultant complexes when small-sized polydentates (usually the bidentates) are involved. As may be easily visualized, the required bending of the small-sized bidentate for its two-site attachment to the positive central metallic atom is accomplished with greater difficulty (if at all) in the *trans* form than in the *cis* form. The distance that must be encompassed by this single ligand in reaching the opposite location sites of the *trans* form is greater than that required to reach the adjacent sites of the *cis* form. This proves to be an experimentally variable matter involving not only the identity of the particular chelating agent, but also the radius ratios and repulsive charges of other ligands that may be present in the complex.

The multi-ringed structures formed by polydentates of higher order — some forming four, five, and six rings — most effectively sequester the central metallic ion. Chelation thus is useful in analytical separations and in industrial processes that seek the removal of deleterious cations. Notable among such processes is water softening by the chelate-complexing of the Ca^{2+}, Mg^{2+}, and Fe^{2+} ions which are responsible for hard water and the boiler scale formed from it.

NOMENCLATURE OF COMPLEXES

The purpose of a systematized procedure for naming coordination compounds is to identify not only their elemental compositions but also, and without ambiguity, the precise chemical constitution of their coordination spheres. Unquestionably, the use here and there of a proprietary or historically inherited name for a particular compound will continue to serve conveniently when there is no doubt as to its chemical associations. The almost bewildering numbers of isomeric coordinated complexes now known, which have been developed since the early work of Werner (1893), make haphazard reference confusingly noninformative.

Scientific rules have been devised for the naming of coordination compounds. The rules apply regardless of where in the compound a complex coordination sphere is found — in anion, or in cation, or, as in some instances, in both.

1. If the coordination compound provides both anion and cation, the cation is always to be named first. For example, if the coordination compound is an uncharged (in net) neutral molecule, its name in general, conforms to that developed for a cation.

COMPLEXING TRANSITIONAL METALS

2. In naming the complexed coordination sphere, ligand anions are to be mentioned *before* ligand molecules. The constituent of the complexed coordination sphere that is to be mentioned *last* is the central metallic ion. If the central metallic atom is present in an anionic coordination sphere, the suffix *ate* is to be appended to the name of the metal, followed thereafter by a parenthesized numeral that designates its oxidation state; thus, (I), (II), (III), (IV), etc. When the oxidation state happens to be zero the Arabic (0) is to be used. If the central metallic ion is in a cationic coordination sphere, or in a neutral complex, its name is not modified — the parenthesized oxidation state of the metallic ion being all that is required.

3. The letter *o* is to be added to the names of all ligand anions in the naming of a complex. Ligand molecules have no characteristic modifications. The derivations shown in Table 3.5 define and illustrate the terminology applicable under this rule.

Table 3.5

Ligand Reference	Non Ligand Reference
acetato-	$C_2H_3O_2^-$, acetate ion
amine	$-NH_2$, amide "group"
ammine	NH_3, ammonia molecule
aquo-	H_2O, water molecule
bromo-	Br^-, bromide ion
carbonato-	CO_3^{2-}, carbonate ion
carbonyl	CO, carbon monoxide molecule
chloro-	Cl^-, chloride ion
cyano-	CN^-, cyanide ion
ethylenediamine	$NH_2CH_2CH_2NH_2$ (= *en*), ethylenediamine molecule
fluoro-	F^-, fluoride ion
hydroxo-	OH^-, hydroxyl ion
iodo-	I^-, iodide ion
nitrato-	NO_3^-, nitrate ion
nitrito-	NO_2^- (= $-ONO^-$), nitrite ion
nitro-	NO_2^- (= $-NO_2^-$), nitrite ion
nitrosyl	NO, nitric oxide molecule
oxalato-	$C_2O_4^{2-}$, oxalate ion
oxo-	O^{2-}, oxide ion
sulfato-	SO_4^{2-}, sulfate ion
sulfido-	S^{2-}, sulfide ion
sulfito-	SO_3^{2-}, sulfite ion
tartrato-	$C_4H_4O_6^{2-}$, tartrate ion
thiocyanato-	SCN^-, thiocyanate ion

4. The number of nonchelate ligands of each chemically different species in the coordination sphere is to be indicated separately by a suitable prefix *di*, *tri*, *tetra*, *penta*, *hexa* (i.e., two, three, four, five, or six — as is applicable). If only one such ligand is present, the prefix *mono* need not be stated; it is understood. When ligands are of the chelate variety, the prefixes to be used to denote their applicable numbers are *bis*, *tris*, and *tetrakis* (twice, thrice, and four times, respectively).

This presentation of the fundamental rules of systematic nomenclature for coordination complexes, may be augmented with additional conventions, but such supplementations would apply only to cases more complicated than those to be presented here.

Some appropriate illustrations of the rules already given are:

$[Co(NH_3)_6]Cl_3 \rightarrow$ hexaamminecobalt(III) chloride
$[Co(NH_3)_6]Cl_2 \rightarrow$ hexaamminecobalt(II) chloride
$K_3[Fe(CN)_6] \rightarrow$ potassium hexacyanoferrate(III)
$K_4[Fe(CN)_6] \rightarrow$ potassium hexacyanoferrate(II).

Note that in each of the preceding compounds (likewise, in those to follow) it is not necessary to provide the species outside of the coordination sphere with prefixes that denote their numbers. The charge upon the coordination sphere itself (automatically the algebraic sum of the individual charges of all the constitutents therein) must, in the completed compound, balance the total electrical charges external to the complex. This fixes unequivocally the number of uncoordinated ions or groups.

Continuing with the illustrations,

$[Co(NH_3)_6]^{3+} \rightarrow$ hexaamminecobalt(III) ion
$[Fe(CN)_6]^{4-} \rightarrow$ hexacyanoferrate(II) ion
$[Cr(NH_3)_3Cl_3]^0 \rightarrow$ trichlorotriamminechromium(III)
$K_4[Ni(CN)_4] \rightarrow$ potassium tetracyanonickel(0)
$[Co(NH_3)_4Cl_2]Br \rightarrow$ dichlorotetraamminecobalt(III) bromide
$[Co(NH_3)_4BrCl]Cl \rightarrow$ bromochlorotetraamminecobalt(III) chloride.

Note that the last two compounds in the preceding list are isomeric and that their respective coordination spheres, however, are not.

Proceeding with the illustrations;

$[Ag(NH_3)_2]_2[PtI_4] \rightarrow$ diamminesilver(I) tetraiodoplatinate(II)
$[Cr(NH_3)_5H_2O][Co(CN)_6] \rightarrow$ aquopentaamminechromium(III) hexacyanocobaltate(III)
$[Pt(NH_3)_4][PtBr_4] \rightarrow$ tetraammineplatinum(II) tetrabromoplatinate(II)
$[Pt(NH_3)_3Br][Pt(NH_3)Br_3] \rightarrow$ bromotriammineplatinum(II) tribromoammineplatinate(II).

Note that of this preceding set of compounds, the last two are isomers; but, as in the set ahead of this, the coordination spheres of these two compounds are not isomerically related. In final illustration,

$[Ni(en)_2]^{2+} \rightarrow$ bis(ethylenediamine)nickel(II) ion
$[Co(en)_2(NH_3)_2]^{3+} \rightarrow$ bis(ethylenediamine)diamminecobalt(III) ion
$[Ni(en)_2]SO_4 \rightarrow$ bis(ethylenediamine)nickel(II) sulfate
$[Pt(en)(NH_3)_4]Cl_4 \rightarrow$ ethylenediaminetetraammineplatinum(IV) chloride.

STABILITY AND LABILITY OF COMPLEXES

The potential concentrations of uncoordinated cations, anions, or molecules that must be anticipated to remain in equilibrium with the coordination complexes that they can form by ligand associations, are prime considerations in the applications to which the complexing process is put in the separation and identification of chemical species. The precipitation, solubilization, ionization, or dissociation that must be either promoted or inhibited for effective analytical procedures becomes very largely involvements of:

(1) the feasibility and predictability of the particular extent of the existence or the destruction of complexes under different chemical conditions; that is, the *stability* of complexes.

(2) the rates at which complexes are formed or destroyed; that is, the *lability* of complexes.

Stability interprets thermodynamically the possibility that a given chemical change will or will not take place; the concept of rate is not involved. Lability, on the other hand, evaluates the kinetic aspect of the relative speed at which the particular chemical change occurs. It is important to understand this distinction, because stability and lability by no means necessarily parallel each other. Just as the most powerful oxidizing agent need not be the one that does the job in the least time, so, likewise, the most stable complex does not, necessarily, take the longest time to dissociate into its component equilibrium simple cations and anions (or molecules).

Thermodynamic stability of a complex is measured by the relative magnitude of the equilibrium constant for a particular chemical change in which it is involved. When the equilibrium constant is represented by $K_{\text{instability}}$, it refers to a chemical process involving the pure aqueous dissociation of the complex into its component species. For example,

$$\text{Co(NH}_3)_6^{3+} + 6\text{H}_2\text{O} \rightleftarrows \text{Co(H}_2\text{O})_6^{3+} + 6\text{NH}_3$$

or, with the solvent omitted, and as typical of the numerical evaluations found for yure water solutions in the usual compendia of instability constants,

$$\text{Co(NH}_3)_6^{3+} \rightleftarrows \text{Co}^{3+} + 6\text{NH}_3,$$

Mass action is then expressed by

$$\frac{[\text{Co}^{3+}][\text{NH}_3]^6}{[\text{Co(NH}_3)_6^{3+}]} = K_{\text{inst}} = 2.2 \times 10^{-34}.$$

Clearly, this very small value for K_{inst} denotes that in pure aqueous medium the cobaltic ammine is thermodynamically highly stable, or merely called stable without additional qualification. On the other hand, when the cobaltic ammine is treated with HCl, the ligand exchange by the resultant reaction,

$$Co(NH_3)_6^{3+} + 6H_3O^+ \rightleftarrows Co(H_2O)_6^{3+} + 6NH_4^+$$

or simply

$$Co(NH_3)_6^{3+} + 6H^+ \rightleftarrows Co^{3+} + 6NH_4^+$$

is evaluated by a new constant corresponding to the following expression for mass action:

$$\frac{[Co^{3+}][NH_4^+]^6}{[Co(NH_3)_6^{3+}][H^+]^6} = K_{equil} = \frac{K_{inst} \times K_b^6 NH_3}{K_{H_2O}^6} = 7.5 \times 10^{21}.$$

The extremely large value determined for $K_{equilibrium}$ of this reaction clearly stresses the thermodynamic instability of the cobaltic ammine complex in an acidic environment. The point to be made is that despite its unstable character in HCl (as contrasted with its high stability in pure water), the cobaltic ammine attains the equilibrium dissociation value expressed for it by $K_{equilibrium}$ with such extreme slowness that for all practical purposes it may be regarded as being virtually *inert;* that is, its lability is extremely low. It is thus possible to maintain the $Co(NH_3)_6^{3+}$ ion in moderately concentrated HCl solution for extensive periods of time without its decomposing to any significant extent.

Interestingly enough, this inertness of the cobalt*ic* ammine, $Co(NH_3)_6^{3+}$, contrasts with the high lability of its +2 analogue — the cobalt*ous* ammine, $Co(NH_3)_6^{2+}$. The latter complex ion attains instantaneous equilibrium with component equilibrium species in all media. The highly labile character of the cobaltous ammine thus permits a prompt and confident prediction of the varying extents to which it will dissociate under variable chemical conditions. This is not readily possible with the cobaltic ammine or, for that matter, with other inert complexes.

Complexes ordinarily prove to be quite labile. Of the fairly common metals, perhaps the most frequently encountered exceptions to easy lability are the complexes of cobalt(III) and of chromium(III).

CF THEORY AND SOME COMPARISONS WITH VALENCE BONDING

The simplicity and convenience of CF concepts of chemical bonding in the complexing of transition metals should not beguile the student away from an ever-necessary awareness that many complexities of coordination require, for successful interpretation, the injection of additional elements of reasoning. Particularly is this true in the quantitative — rather than qualitative — evaluation of complexing processes. The CF theory views complexing as a strictly ionic or ion-dipole interaction of electrostatic attractions between the centrally situated metallic ion and the ligands of its coordination sphere. There is, however, ample evidence in nearly all cases of complexing of limited degrees of covalent bonding as well. Covalency in bonding, expressive of the overlap of atomic orbitals, represents the outer extreme of explanatory concept, to which both the molecular orbital and the valence bond theories defer, and with which the ligand field theory attempts to compromise. Nonetheless, the purely qualitative electrostatics of the CF theory, without the complications of overlap, readily serve the present objectives.

CRYSTAL FIELD SPLITTING IN OCTAHEDRAL COMPLEXES

Recall the geometric distribution of the charged clouds of d orbitals with relation to their Cartesian axes of symmetry (x, y, and z). The d_{xy}, d_{xz}, and d_{yz} orbitals are oriented *between* the axes of symmetry, the d_{z^2} and $d_{x^2-y^2}$ orbitals are oriented *along* these axes. In any given crystal of a complex compound, the distribution of the electron clouds that determine the symmetry of the coordination sphere may be qualitatively generalized as the interplay and counterplay of two operating forces:

(1) the mutual attractions between a single electropositive central cation and the electronegative anions or negative ends of dipole molecules representing the ligands. These forces in themselves clearly operate to draw the species together as closely as possible.

(2) the mutual repulsions between and among the clouds of electrons of the ligands themselves. Clearly, these operate to push the ligands away from one another as widely as possible.

These two opposing forces must be operationally compromised. The very widest separation possible among the six ligands in an octahedral coordination sphere, as they approach the metallic ion centralized at the origin of the three Cartesian axes x, y, and z, is on the direct lines of these axes, with the ligands entering into the coordination sphere from both ends of each axis simultaneously (Fig. 3.5).

Visualize, now, the orientation of the electron clouds in the set of five d orbitals that belong to the metallic ion. Oriented as they are, along the Cartesian coordinate axes, the lobes of the d_{z^2} and $d_{x^2-y^2}$ orbitals are consequently directly in the path of the oncoming ligands. The electrons in these orbitals therefore are subjected to greater forces of repulsion from the electron clouds of the ligands than those in the d_{xy}, d_{xz}, and d_{yz} orbitals. These latter three orbitals are in positions predominantly lateral to, or

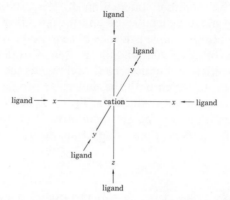

FIG. 3.5 Central metallic ion and ligands in octahedral coordination sphere.

above and below, the electron clouds of the approaching ligands; that is, the ligands approach between the lobes of these orbitals. In the free, uncomplexed metallic ion, all the d orbitals are energetically equivalent — so-called "degenerate" orbitals.

This means that an electron can readily occupy with equal preference any one of the five orbitals. The very much stronger repulsion to which an electron in a d_{z^2} or $d_{x^2-y^2}$ orbital is subjected, as contrasted with one in any of the d_{xy}, d_{xz}, or d_{yz} orbitals, compels its being displaced to an orbital of greater stability or lower energy. Hence, under the influence of an approaching ligand, the five d orbitals that, originally, were all energetically equivalent in the free, uncomplexed metallic cation, now become, in the complex, split into two separate sublevels of differing energy. In the higher energy level of the octahedral crystal (conventionally symbolized as the "eg" energy level), consequently, is found the d_{z^2} and $d_{x^2-y^2}$ orbitals; in the lower energy level (conventionally symbolized as the "t_{2g}" energy level) is found the d_{xy}, d_{xz}, and d_{yz} orbitals. This separation of the free, uncomplexed set of five orbitals, hitherto energetically equivalent, into orbital subgroups of different sets of energies, is known as *crystal field splitting*.

The extent of crystal field splitting must be evaluated for a complex first in terms of the charge upon a central metallic ion. The greater the charge, the closer the ligands are to the central metallic ion, and the greater is the splitting. This must also be evaluated in terms of the repulsions exerted by the ligands on the crystal field. The greater their repulsive effects, the greater the splitting. On a strictly relative basis, however, the extent of d-orbital separation is designated as "strong field" and "weak field."

To illustrate strong-field and weak-field splitting, the cyanide ion (CN^-), in complexing, invariably exerts a strong field; the halide ions (F^-, Cl^-, Br^-, I^-), exert relatively weak fields. Comparisons such as these, however, must be made quite guardedly, because a completely plausible evaluation of the relative strength of crystal fields cannot ignore the charge or size of the central metallic ion involved. When qualitatively compared by complexing with the identical cation, the crystal field splitting exerted by the following frequently encountered ligands usually increases in the following order:

$$I^- < Br^- < Cl^- < F^- < OH^- < H_2O < NH_3 < CN^-$$

Increasing strengths of the respective crystal fields →

In pondering Figure 3.6 which illustrates the foregoing concepts, it is to be observed that the repulsions of the crystal field raise the energy levels of all the orbitals of the hitherto uncomplexed cation. This is a concomitant adjunct of the splitting that characterizes the crystal field of the complex. Note, not only that ΔE, representing the energy difference of separation between the two sets of orbitals split in the crystal field, is smaller in the weak field complex than in the strong field; but also that both the e_g and the t_{2g} levels of the weak field are correspondingly closer to the energies of the degenerate orbitals of the free isolated metallic ion than are those, respectively, of the strong field. The limited objectives here in the development of the electrostatic crystal field theory are amply fulfilled by forthcoming specific illustrations in octahedral 6-coordination complexing. Meanwhile, Figure 3.6 (a) reveals the similarities and differences in the applications of this theory to the formation of complexes in 4-coordination, both tetrahedral and square planar.

CRYSTAL FIELD DELINEATIONS OF CERTAIN 6-COORDINATION COMPLEXES

In the following examples of octahedral crystal field concepts we pursue the adjustments in orbital locations made by electrons in response to repulsions by ligands. In all cases, the theory of the justification of the pertinent adjustment conform to measured magnetic moments, and to the determinable extent of paramagnetism of the specific species. A few guides, if borne in mind, lead to a correct, simple, and convenient interpretation of the electron displacements that must parallel any alterations in the magnetic properties of the free simple metallic ion, which are induced as a result of its being complexed, and which must likewise conform to the magnetic properties of the complex itself:

1. The s and p electrons of the transition metal are not involved in the complexing process.

Were they to be involved, the relative strength of the crystal field and the pertinent value of its ΔE (representative of d-orbital splitting) would not be significantly affected, inasmuch as all s and p electrons sustain fairly equal repulsions by ligands. As the three p orbitals are not split, and as the

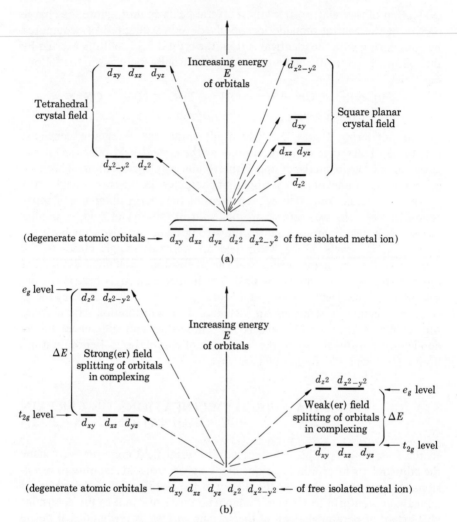

FIG. 3.6 Energy levels of crystal field splitting. (a) 4-Tetrahedral and square planar. (b) 6-coordination (octahedral) splitting. Strong and weak fields.

single s orbital is completely immune to any divisive influences with respect to its total energy, the s and p electrons in the octahedral electrostatic field of a transitional-type complex are without influence in altering the magnetic properties of the coordinated species. This contrasts with those of the free uncomplexed metallic ion from which it is derived. This consideration, however, must not be confused with an awareness that unpaired electrons in any orbital, be it s, p, d, or f, contribute to the over-all paramagnetic behavior of the compound as a whole.

2. Electrons tend to remain unpaired as long as possible (Hund's Rule of Maximum Multiplicity).

Only when separate orbitals are no longer available to accommodate single electrons of parallel spin, do electrons pair in the individual orbitals in opposing spin. The spin-paired condition is aided by a strong-field ligand which, by its greater repulsive force upon d electrons in the cation, compels their mutual associations in the lower energy level of the octahedral crystal field; that is, the t_{2g} level of d_{xy}, d_{xz}, and d_{yz} orbitals.

A weak-field ligand in complexing tends to retain d-orbital electrons in a spin-unpaired condition, which will result in comparatively little or no change in the d-orbital distributions of electrons of the uncomplexed cation. In the weak field, consequently, the higher e_g level of d_{z^2} and $d_{x^2-y^2}$ orbitals proves almost as favorable for occupation by electrons, as does the lower t_{2g} level. These are extremes, but moderate crystal fields may be appraised with the same plausible approaches.

3. Electrons tend to occupy orbitals of lowest energy, and greatest stability. Only when they lack accommodation in the more stable, less energetic lower level (t_{2g}) of the d_{xy}, d_{xz}, and d_{yz} orbitals of the octahedral crystal field do electrons enter the higher level (e_g) of d_{z^2} and $d_{x^2-y^2}$ orbitals.

With the foregoing explanations understood, it should be possible to interpret Figure 3.7 and Table 3.6 without difficulty.

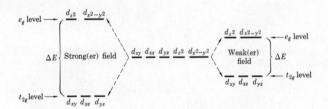

FIG. 3.7 Algebraic mean of complexing orbitals of metallic ion in energy states unable to differentiate between upper and lower levels.

A few items respecting Table 3.6 are deserving of mention.

1. The exclusions from the delineations given of a d^0 ion, such as Sc(III), and of a d^{10} ion such as Zn(II) have been deliberate. Crystal field expositions herein contend with the electrostatics of the presence of one or more d electrons in the free uncomplexed metal ion, and upon the availability of empty or partially-filled orbitals in the sets of differing energy levels. In Sc(III), there are no d electrons; hence it is not a transitional species, as in contrast the neutral scandium atom, Sc^0, which has a single d electron, and which, consequently, is transitional. In Zn(II), all d orbitals of the free metal ion are complete with the maximum possible five sets of spin-paired

104 COMPLEXING TRANSITIONAL METALS

Table 3.6   Distribution of electrons in octahedral disposition

Species	strong-field complex	uncomplexed metal ion	weak-field complex
d^1	$(e_g)^0$ — — $(t_{2g})^1$ ↑ _ _ unpaired = 1	↑ _ _ _ _ (e.g., Ti^{III}, V^{IV}) 1	— — $(e_g)^0$ ↑ _ _ $(t_{2g})^1$ 1
d^2	$(e_g)^0$ — — $(t_{2g})^2$ ↑ ↑ _ unpaired = 2	↑ ↑ _ _ _ (e.g., Ti^{II}, V^{III}) 2	— — $(e_g)^0$ ↑ ↑ _ $(t_{2g})^2$ 2
d^3	$(e_g)^0$ — — $(t_{2g})^3$ ↑ ↑ ↑ unpaired = 3	↑ ↑ ↑ _ _ (e.g., V^{II}, Cr^{III}, Mo^{III}) 3	— — $(e_g)^0$ ↑ ↑ ↑ $(t_{2g})^3$ 3
d^4	$(e_g)^0$ — — $(t_{2g})^4$ ↑↓ ↑ ↑ unpaired = 2	↑ ↑ ↑ ↑ _ (e.g., Cr^{II}, Mn^{III}) 4	↑ _ $(e_g)^1$ ↑ ↑ ↑ $(t_{2g})^3$ 4
d^5	$(e_g)^0$ — — $(t_{2g})^5$ ↑↓ ↑↓ ↑ unpaired = 1	↑ ↑ ↑ ↑ ↑ (e.g., Fe^{III}, Mn^{II}) 5	↑ ↑ $(e_g)^2$ ↑ ↑ ↑ $(t_{2g})^3$ 5
d^6	$(e_g)^0$ — — $(t_{2g})^6$ ↑↓ ↑↓ ↑↓ unpaired = 0	↑↓ ↑ ↑ ↑ ↑ (e.g., Fe^{II}, Co^{III}, Rh^{III}, Ir^{III}, Pt^{IV}) 4	↑ ↑ $(e_g)^2$ ↑↓ ↑ ↑ $(t_{2g})^4$ 4
d^7	$(e_g)^1$ ↑ _ $(t_{2g})^6$ ↑↓ ↑↓ ↑↓ unpaired = 1	↑↓ ↑↓ ↑ ↑ ↑ (e.g., Co^{II}, Ni^{III}) 3	↑ ↑ $(e_g)^2$ ↑↓ ↑↓ ↑ $(t_{2g})^5$ 3
d^8	$(e_g)^2$ ↑ ↑ $(t_{2g})^6$ ↑↓ ↑↓ ↑↓ unpaired = 2	↑↓ ↑↓ ↑↓ ↑ ↑ (e.g., Ni^{II}) 2	↑ ↑ $(e_g)^2$ ↑↓ ↑↓ ↑↓ $(t_{2g})^6$ 2
d^9	$(e_g)^3$ ↑↓ ↑ $(t_{2g})^6$ ↑↓ ↑↓ ↑↓ unpaired = 1	↑↓ ↑↓ ↑↓ ↑↓ ↑ (e.g., Cu^{II}) 1	↑↓ ↑ $(e_g)^3$ ↑↓ ↑↓ ↑↓ $(t_{2g})^6$ 1

electrons; hence, with this ion the orbital splitting characteristic of octahedral field coordination is precluded. In fact, the chemistry of Zn(II) in complexing is strictly that of tetrahedral coordination.

2. The numbers of unpaired electrons do not always serve to distinguish between strong-field and weak-field complexes. In d^4, d^5, d^6, and d^7 species, as the magnetic criteria are different for the two types of complexes they consequently yield the necessary distinctions. In the d^1, d^2, d^3, d^8, and d^9 species, however, the magnetic moments of both strong and weak fields remain unchanged from that of the free uncomplexed transition metallic ion; hence, must be they differentiated by other means, such as examination of their respective spectra.

3. The rather controversial Cu(II) complexes — all of which are d^9 species — are given *octahedral* configurations in their respective crystal fields. This has been a bone of contention between adherents of the valence-bond and the crystal-field theories, inasmuch as the valence-bond concept familarly depicts copper(II) complexes in 4-coordination *square planar* arrangement. This sharply conflicts with the conclusions of crystal-field bonding that copper(II) complexes are merely distorted octahedra.

DISTORTIONS IN CRYSTAL FIELDS

The regular octahedral shape for a complex requires that the centrally situated metallic ion of the coordination sphere have full spherical symmetry. If this is lacking, the proximity of the ligand electron clouds to the center of coordination of the metallic ion — that is, their internuclear distances — cannot possibly be the same for all the ligands. These differences in bond lengths make for distortions of the crystal as a whole, inasmuch as repulsions between ligand electron clouds and/or their attraction to the metallic-ion nucleus cannot then all be similar.

Let us consider the imbalance of the Cu(II) ion with reference to the crystal field already depicted for it as a d^9 species; that is $(t_{2g})^6(e_g)^3$. For either the strong-field or the weak-field complex, the e_g electrons must be represented in their particular arrangements as either

$$(e_g)^3 = (d_{x^2-y^2})^1(d_{z^2})^2$$

or

$$(e_g)^3 = (d_{x^2-y^2})^2(d_{z^2})^1.$$

As a d^{10} ion, e.g., $(t_{2g})^6(e_g)^4$, shows complete spherical symmetry, it would appear that the *absence* of this additional electron in the d^9 copper(II) species is accountable for the latter's distortion in an octahedral field. Let us assume that the e_g arrangement for the copper(II) complex is $(d_{x^2-y^2})^1$

$(d_{z^2})^2$. Inasmuch as the electron clouds of the two orbitals are concentrated along different axes, the absence of a second electron from the x and y planes of the $d_{x^2-y^2}$ orbital removes some of the electron-repulsion shielding that would otherwise compel the electronegative ligand in these planes to keep its distance from the electropositive metallic ion. Consequently, a ligand in these planes may now approach more closely to the central metallic ion than would one situated along the z axis. The result is, thus, the formation of a distorted octahedron constituted of six ligand bonds to the central metallic ion. Two of these bonds are relatively long, and four are shorter and coplanar with the centralized metal. If, instead, the absent electron were assigned to the d_{z^2} orbital rather than to the $d_{x^2-y^2}$ orbital, as in the foregoing illustration, the reverse of this situation could be anticipated. That is, four relatively long bonds would be coplanar with the central metal and there would be two shorter bonds.

Electrostatic crystal fields offer no readily discernible clues to which of the two types of octahedral distortions — the $(d_{x^2-y^2})^1 (d_{z^2})^2$ or the $(d_{x^2-y^2})^2 (d_{z^2})^1$ — may theoretically be predicted to yield the more stable complex. *Experimental* encounters with d^9 crystal line forms are, invariably, bonds of the four-short and one-long type; that is, again, $(d_{x^2-y^2})^1 (d_{z^2})^2$.

In the light of the preceding, a plausible elucidation would suggest that any complex with an e_g level that contains an odd number of electrons should be unable to exist in a regular octahedral field. This checks out quite correctly with the weak-field complexes of the d^4 species. Examples are Cr(II) and Mn(III), which are orbital-split into $(t_{2g})^3 (e_g)^1$; and, likewise, with the strong-field complexes of the d^7 species, Co(II) and Ni(III), which are orbital-split into $(t_{2g})^6 (e_g)^1$. As with the representative species of d^9 complexes, all these yield distorted octahedra.

CERTAIN CONTRASTS IN CF AND VB DELINEATIONS OF COMPLEXING

Careful experimental evaluations of x-ray crystallography and the spectra of complexes yield the desired physical evidences of their geometric configurations and magnetic moments. Investigations of the stability of complexes likewise provide informative results. All these, necessarily, must be the basis for any theoretical postulates that are to be applied to an imaginative interpretation of the bondings of the d orbitals. Even the numbers of spin-paired electrons which prove so useful as guides to the strengths of electrostatic ligand fields and predictions of bonding are, themselves, merely theoretical elucidations derived from an experimental quantity, the *magnetic moment*. The interplay and counterplay of extremes of theoretical concept that attempt to interpret the bondings of orbitals and their energy transformations in complexing must, more often than not, be tempered by compro-

COMPLEXING TRANSITIONAL METALS   107

mise, and by a keen sense of awareness that, in the absence of all the facts, no theory can possibly provide all the answers.

The pure electrostatics of crystal field complexing are now to be contrasted with the valence bond concept of hybridization or union of orbitals. Valence-bonding in a complex may be viewed as the accommodation within a vacant atomic orbital of the central metallic ion, of a set of spin-paired electrons that are associated with a filled atomic orbital of a ligand. The overlapping of the respective orbitals of the ligand and central metallic ion required to accomplish this, results in the conversion of certain of the originally non-equivalent orbitals of the cation to hybrid orbitals of presumed equivalency of energy. The interpretive comparisons follow:

1.   A d^5 species derived from $\begin{Bmatrix} Fe^{3+} \\ Mn^{2+} \end{Bmatrix}$ or $\rightarrow$

$\qquad\qquad\qquad\qquad\quad\uparrow\quad\uparrow\quad\uparrow\quad\uparrow\quad\uparrow$
(Ar core)^{18} $3d'$ $3d'$ $3d'$ $3d'$ $3d'$ $4s^0$ $4p^0$ $4p^0$ $4p^0$

(a)   The strong crystal-field effects of 6 CN$^-$ ligands on the d orbitals of the cation:

[Fe(CN)$_6$]$^{3-}$
or $\cdots\cdots$ $\begin{bmatrix} d(e_g)^0 \text{ — —} \\ d(t_{2g})^5 \; \uparrow\downarrow \; \uparrow\downarrow \; \uparrow \end{bmatrix}$ 1 unpaired electron.
[Mn(CN)$_6$]$^{4-}$

(b)   Valence-bonding of six CN$^-$ ligands to the central metallic ion to form an *inner-orbital* complex.

This is so called in valence-bond terminology because it utilizes d orbitals in a level *below* that of the principal quantum level of the bonding s and p orbitals of the metallic ion.

In order to produce the experimental paramagnetic effect corresponding to one unpaired electron, as determined, the electrons in the d orbitals must be regrouped by pairing. This provides the full complement of six vacant orbitals required to accommodate the ligands. The d^2sp^3 hybrid thus formed, which means the union of two d, one s, and three p orbitals of the metallic ion, always results in an octahedral configuration of the complex.

[Fe(CN)$_6$]$^{3-}$ available for inner-orbital
                                                 bonding with 6 CN$^-$
or $\cdots\cdots$ — — — $\overbrace{\text{— — — — — —}}$ (1 unpaired electron)
[Mn(CN)$_6$]$^{4-}$ $3d^2$ $3d^2$ $3d^1$ $\underbrace{3d^0\;3d^0\;4s^0\;4p^0\;4p^0\;4p^0}_{d^2sp^3 \text{ hybridization}}$

2.   A d^5 species derived from
$\qquad$ Fe$^{3+}\rightarrow$ (Ar core)^{18} $\;\uparrow\quad\uparrow\quad\uparrow\quad\uparrow\quad\uparrow$
$\qquad\qquad\qquad\qquad\qquad\quad 3d^1\;3d^1\;3d^1\;3d^1\;3d^1\;4s^0\;4p^0\;4p^0\;4p^0$

(a) The weak crystal-field effects of 6 F^- ions or 6 H_2O dipoles on the d orbitals of Fe^{3+}:

$[FeF_6]^{3-}$
or ..... $\begin{bmatrix} d(e_g)^2 & \uparrow & \uparrow \\ d(t_{2g})^3 & \uparrow & \uparrow & \uparrow \end{bmatrix}$ 5 unpaired electrons.
$[Fe(H_2O)_6]^{3+}$

(b) Valence-bonding of six F^- or six H_2O ligands to form *outer-orbital* complex.

In VB terminology this indicates the utilization of d orbitals of the metallic ion on the *same* principal quantum level as that of its bonding s and p orbitals.

The five unpaired electrons that correspond to the measured magnetic moments of $[FeF_6]^{3-}$ and $[Fe(H_2O)_6]^{3+}$ require that the spin-unpaired assignments of the free cation be maintained. Required, also, are six empty orbitals for bonding with six F^- ligands. The utilization of two of the five $4d$ orbitals is consequently indicated in the valence-bond concept of the complex to be formed. This sp^3d^2 hybrid signifies the employment of *one s*, *three p*, and *two d* orbitals of the metallic ion. It always results in an octahedral configuration of the complex.

$[FeF_6]^{3-}$
or ..... $\uparrow \quad \uparrow \quad \uparrow \quad \uparrow \quad \uparrow$
$[Fe(H_2O)_6]^{3+}$ $\quad 3d^1 \; 3d^1 \; 3d^1 \; 3d^1 \; 3d^1$

$\underbrace{\qquad \qquad \qquad \qquad \qquad}_{\text{available for outer-orbital bonding with 6 } F^- \text{ or 6 } H_2O}$

$\underbrace{4s^0 \; 4p^0 \; 4p^0 \; 4p^0 \; 4d^0 \; 4d^0 \; 4d^0 \; 4d^0 \; 4d^0}_{sp^3d^2 \text{ hybridization}}$ (5 unpaired electrons)

3. A d^6 species derived from $\begin{Bmatrix} Fe^{2+} \\ Co^{3+} \end{Bmatrix}$ or $\rightarrow$

(Ar core)^{18} $\uparrow\downarrow \; \uparrow \; \uparrow \; \uparrow \; \uparrow$
$\qquad \qquad \quad 3d^2 \; 3d^1 \; 3d^1 \; 3d^1 \; 3d^1 \; 4s^0 \; 4p^0 \; 4p^0 \; 4p^0$

(a) A strong crystal field:

$[Fe(CN)_6]^{4-}$
or ..... $\begin{bmatrix} d(e_g)^0 & - & - \\ d(t_{2g})^6 & \uparrow\downarrow & \uparrow\downarrow & \uparrow\downarrow \end{bmatrix}$ 0 unpaired electrons
$[Co(CN)_6]^{3-}$ $\qquad \qquad \qquad \qquad \qquad \qquad \qquad$ *diamagnetic*

(b) A valence-bonded inner-orbital complex:

$[Fe(CN)_6]^{4-}$ $\qquad \qquad$ $\overbrace{\qquad \qquad \qquad}^{\text{available for inner-orbital bonding with 6 } CN^-}$

or ..... $\uparrow\downarrow \quad \uparrow\downarrow \quad \uparrow\downarrow$ $\qquad \qquad \qquad \qquad$ (0 unpaired electrons)
$\qquad \qquad \; 3d^2 \; 3d^2 \; 3d^2 \; \underbrace{3d^0 \; 3d^0 \; 4s^0 \; 4p^0 \; 4p^0 \; 4p^0}_{d^2sp^3 \text{ hybridization}}$ *diamagnetic*
$[Co(CN)_6]^{3-}$

COMPLEXING TRANSITIONAL METALS 109

Note the regrouping by pairing, of electrons in the $3d$ level in order to provide the necessary number of empty orbitals for ligand-bonding.

(c) A weak crystal field:

$$[CoF_6]^{3-} \cdots \cdots \begin{bmatrix} d(e_g)^2 & \uparrow & \uparrow & \\ d(t_{2g})^4 & \uparrow\downarrow & \uparrow & \uparrow \end{bmatrix} \text{4 unpaired electrons}$$

(d) A valence-bonded outer-orbital complex:

$[CoF_6]^{3-} \cdots \cdots$ — — — — —
$\phantom{[CoF_6]^{3-} \cdots}$ $3d^2\ 3d^1\ 3d^1\ 3d^1\ 3d^1$
$\phantom{[CoF_6]^{3-} \cdots}$ available for outer-orbital
$\phantom{[CoF_6]^{3-} \cdots}$ bonding with 6 F⁻

$\underbrace{\ \_\ \_\ \_\ \_\ \_\ \_\ \_\ \_\ \_\ }$ (4 unpaired electrons)
$4s^0\ 4p^0\ 4p^0\ 4p^0\ 4d^0\ 4d^0\ 4d^0\ 4d^0\ 4d^0$
sp^3d^2 hybridization

4. A d^9 species derived from
 Cu²⁺ → (Ar core)¹⁸ $\uparrow\downarrow\ \uparrow\downarrow\ \uparrow\downarrow\ \uparrow\downarrow\ \uparrow$
 $3d^2\ 3d^2\ 3d^2\ 3d^2\ 3d^1\ 4s^0\ 4p^0\ 4p^0\ 4p^0$

(a) Strong and weak crystal fields yield identical distribution of electrons in the orbital-split levels:

$$[Cu(NH_3)_4]^{2+} \cdots \cdots \begin{bmatrix} d(e_g)^3 & \uparrow\downarrow & \uparrow & \\ d(t_{2g})^6 & \uparrow\downarrow & \uparrow\downarrow & \uparrow\downarrow \end{bmatrix} \text{1 unpaired electron}$$

(b) valence-bonding requiring the "promotion" of an electron.

With the paramagnetism of the copper(II) ammine complex experimentally determined to correspond to one unpaired electron (also true of the free Cu²⁺ ion), "dilemma" in valence-bonding is encountered with respect to the availability of requisite bonding orbitals.

The valence-bond concept represents the Cu(II) complexes (e.g., Cu(NH₃)₄²⁺) as square planar. This is in sharp disagreement with crystal-field concepts which characterize the Cu(NH₃)₄²⁺ complex as a *distorted octahedron* — in valid conformity with its spectral properties. In its present conformity to the square planar configuration assigned to this complex ion, the valence-bond theory necessarily precludes the utilization of the one $4s$ and the three $4p$ orbitals that are vacant in the structure of the Cu²⁺ ion, inasmuch as this would result in a totally unconforming tetrahedral sp^3 bonding in the formation of the complex.

As square planar 4-coordination bonding is systematized as dsp^2 hybridization, valence-bonding resolves its difficulty of finding the necessary completely vacant d orbital by promoting the unpaired electron from its $3d$ orbital to a $4p$ orbital; thus

$$\underset{3d^2\ 3d^2\ 3d^2\ 3d^2}{\uparrow\downarrow\ \uparrow\downarrow\ \uparrow\downarrow\ \uparrow\downarrow}\ \underbrace{\overbrace{3d^0\ 4s^0\ 4p^0\ 4p^0\ 4p^1}^{\text{available for bonding}}}_{dsp^2\ \text{hybridization}}\quad \text{(1 unpaired electron)}$$

ADDITIONAL ILLUSTRATIONS OF VALENCE-BONDING

These additional illustrations are presented to provide serviceable methods of delineating tetrahedral, square planar, and linear arrangements of coordination.

A d^{10} species derived from $Zn^{2+} \rightarrow$

$$(\text{Ar core})^{18}\ \underset{3d^2\ 3d^2\ 3d^2\ 3d^2\ 3d^2\ 4s^0\ 4p^0\ 4p^0\ 4p^0}{\uparrow\downarrow\ \uparrow\downarrow\ \uparrow\downarrow\ \uparrow\downarrow\ \uparrow\downarrow}$$

There are no unpaired electrons in a d^{10} species; and no $3d$ orbitals either to utilize or to vacate for bondings of Zn^{2+} with ligands. Such familiar complexes as $Zn(NH_3)_4^{2+}$, $Zn(OH)_4^{2-}$, and $Zn(CN)_4^{2-}$ yield the tetrahedral 4-coordination arrangement, exclusively. Hybridized for this purpose are the single $4s$ and the three $4p$ orbitals of the Zn^{2+} ion. This produces the sp^3 hybrid always proprietary to the tetrahedral configuration.

$$\underset{3d^2\ 3d^2\ 3d^2\ 3d^2\ 3d^2}{\uparrow\downarrow\ \uparrow\downarrow\ \uparrow\downarrow\ \uparrow\downarrow\ \uparrow\downarrow}\ \underbrace{\overbrace{4s^0\ 4p^0\ 4p^0\ 4p^0}^{\text{available for bonding}}}_{sp^3\ \text{hybridization}}\quad \begin{matrix}\text{(0 unpaired electrons)}\\ \textit{diamagnetic}\end{matrix}$$

Square Planar Arrangement

Square planar arrangement presents complexing of the same metallic ion in an identical coordination number of more than one type of spatial geometry arrangement. Typical of such stereochemical adaptability to multiple geometric forms, is the Ni^{2+} ion. This simple cation yields 4-coordination complexes in both square planar and tetrahedral arrangements. For any specific complex, however, not only must the coordination number of the metal ion be constant, but also its geometric configuration. The structural duality displayed by the Ni^{2+} ion in 4-coordination forms depends upon the nature of the ligands. Thus, $Ni(H_2O)_4^{2+}$ is tetrahedral, whereas $Ni(CN)_4^{2-}$ is square planar.

Ni^{2+}(a d^8 species) $\rightarrow (\text{Ar core})^{18}\ \underset{3d^2\ 3d^2\ 3d^2\ 3d^1\ 3d^1\ 4s^0\ 4p^0\ 4p^0\ 4p^0}{\uparrow\downarrow\ \uparrow\downarrow\ \uparrow\downarrow\ \uparrow\ \uparrow}$

COMPLEXING TRANSITIONAL METALS

$[\text{Ni}(\text{H}_2\text{O})_4]^{2+}\cdots\cdots$ $\underbrace{\underset{3d^2}{\uparrow\downarrow}\ \underset{3d^2}{\uparrow\downarrow}\ \underset{3d^2}{\uparrow\downarrow}\ \underset{3d^1}{\uparrow}\ \underset{3d^1}{\uparrow}}\ \overbrace{\underset{4s^0}{}\ \underset{4p^0}{}\ \underset{4p^0}{}\ \underset{4p^0}{}}^{\text{available for bonding}}$ (2 unpaired electrons)

sp^3 hybridization
(tetrahedral)

$[\text{Ni}(\text{CN})_4]^{2-}\cdots\cdots$ $\underset{3d^2}{\uparrow\downarrow}\ \underset{3d^2}{\uparrow\downarrow}\ \underset{3d^2}{\uparrow\downarrow}\ \underset{3d^1}{\uparrow\downarrow}\ \overbrace{\underset{3d^0}{}\ \underset{4s^0}{}\ \underset{4p^0}{}\ \underset{4p^0}{}\ \underset{4p^0}{}}^{\text{available for bonding}}$ (0 unpaired electrons)
diamagnetic

dsp^2 hybridization
(square planar)

Note that for the $\text{Ni}(\text{CN})_4^{2-}$ ion, the regrouping by the spin-pairing of electrons of the simple uncomplexed Ni^{2+} ion reduces the paramagnetism of the latter to zero and makes available the single d orbital required for the square planar (dsp^2) arrangement of the complex.

The Ni^{2+} ion can also form 6-coordination octahedral complexes, i.e., $\text{Ni}(\text{NH}_3)_6^{2+}$, an sp^3d^2 outer-orbital complex.

Linear Arrangement

Linear arrangement is exclusive to 2-coordination complexes and always results in sp hybrids.

A d^{10} species derived from
$$\text{Ag}^+ \rightarrow (\text{Kr core})^{36}\ \underset{4d^2}{\uparrow\downarrow}\ \underset{4d^2}{\uparrow\downarrow}\ \underset{4d^2}{\uparrow\downarrow}\ \underset{4d^2}{\uparrow\downarrow}\ \underset{4d^2}{\uparrow\downarrow}\ \underset{5s^0}{}\ \underset{5p^0}{}\ \underset{5p^0}{}\ \underset{5p^0}{}$$

$\text{Ag}(\text{NH}_3)_2^+$
or $\cdots\cdots$ $\underset{4d^2}{\uparrow\downarrow}\ \underset{4d^2}{\uparrow\downarrow}\ \underset{4d^2}{\uparrow\downarrow}\ \underset{4d^2}{\uparrow\downarrow}\ \underset{4d^2}{\uparrow\downarrow}\ \overbrace{\underset{5s^0}{}\ \underset{5p^0}{}\ \underset{5p^0}{}\ \underset{5p^0}{}}^{\text{available for bonding}}$ (0 unpaired electrons)
$\text{Ag}(\text{CN})_2^-$ *diamagnetic*

sp hybridization
(linear)

TETRAHEDRAL AND SQUARE PLANAR ELECTROSTATIC CRYSTAL FIELDS

The crystal fields, and extent of paramagnetism of complexes in tetrahedral and square planar arrangements may be interpreted by applying the same considerations conceived for the preferential filling of the split orbitals in octahedral orientations. These are fundamentally the separate maintenance of unpaired electrons of parallel spin in individually separate orbitals wherever possible (Hund's rule). In opposition to this is the untenability of such separate existence when, in the presence of strong ligand fields, lower energy levels of greater stability are available.

Table 3.7

Crystal fields

STRONG TETRAHEDRAL

Number of d electrons	d^1	d^2	d^3	d^4	d^5	d^6	d^7	d^8	d^9
$d_{xy}\ d_{xz}\ d_{yz}$ $\left[\begin{array}{c}d\text{-orbital}\\ distributions\end{array}\right]$ $d_{x^2-y^2}\ d_{z^2}$	— — — $\underline{\uparrow}\ —$	— — — $\underline{\uparrow}\ \underline{\uparrow}$	— — — $\underline{\uparrow\downarrow}\ \underline{\uparrow}$	— — — $\underline{\uparrow\downarrow}\ \underline{\uparrow\downarrow}$	$\underline{\uparrow}\ —\ —$ $\underline{\uparrow\downarrow}\ \underline{\uparrow\downarrow}$	$\underline{\uparrow}\ \underline{\uparrow}\ —$ $\underline{\uparrow\downarrow}\ \underline{\uparrow\downarrow}$	$\underline{\uparrow}\ \underline{\uparrow}\ \underline{\uparrow}$ $\underline{\uparrow\downarrow}\ \underline{\uparrow\downarrow}$	$\underline{\uparrow\downarrow}\ \underline{\uparrow}\ \underline{\uparrow}$ $\underline{\uparrow\downarrow}\ \underline{\uparrow\downarrow}$	$\underline{\uparrow\downarrow}\ \underline{\uparrow\downarrow}\ \underline{\uparrow}$ $\underline{\uparrow\downarrow}\ \underline{\uparrow\downarrow}$ energy E increases
Unpaired electrons	1	2	1	0	1	2	3	2	1

WEAK TETRAHEDRAL

Number of d electrons	d^1	d^2	d^3	d^4	d^5	d^6	d^7	d^8	d^9
$d_{xy}\ d_{xz}\ d_{yz}$ $\left[\begin{array}{c}d\text{-orbital}\\ distributions\end{array}\right]$ $d_{x^2-y^2}\ d_{z^2}$	— — — $\underline{\uparrow}\ —$	— — — $\underline{\uparrow}\ \underline{\uparrow}$	$\underline{\uparrow}\ —\ —$ $\underline{\uparrow}\ \underline{\uparrow}$	$\underline{\uparrow}\ \underline{\uparrow}\ —$ $\underline{\uparrow}\ \underline{\uparrow}$	$\underline{\uparrow}\ \underline{\uparrow}\ \underline{\uparrow}$ $\underline{\uparrow}\ \underline{\uparrow}$	$\underline{\uparrow\downarrow}\ \underline{\uparrow}\ \underline{\uparrow}$ $\underline{\uparrow}\ \underline{\uparrow}$	$\underline{\uparrow\downarrow}\ \underline{\uparrow\downarrow}\ \underline{\uparrow}$ $\underline{\uparrow}\ \underline{\uparrow}$	$\underline{\uparrow\downarrow}\ \underline{\uparrow\downarrow}\ \underline{\uparrow}$ $\underline{\uparrow\downarrow}\ \underline{\uparrow}$	$\underline{\uparrow\downarrow}\ \underline{\uparrow\downarrow}\ \underline{\uparrow}$ $\underline{\uparrow\downarrow}\ \underline{\uparrow\downarrow}$ energy E increases
Unpaired electrons	1	2	3	4	5	4	3	2	1

STRONG SQUARE PLANAR

Number of d electrons	d^1	d^2	d^3	d^4	d^5	d^6	d^7	d^8	d^9
d electrons distributions $\overline{d_{x^2-y^2}}$ $\overline{d_{xy}}$ $\overline{d_{xz}}\,\overline{d_{yz}}$ $\overline{d_{z^2}}$	— — ↑ — —	— ↑↓ — —	— ↑↓ ↑ ↑	— ↑↓ ↑↓ ↑	— ↑↓ ↑↓ ↑↓ ↑	— ↑↓ ↑↓ ↑↓ ↑↓	— ↑↓ ↑↓ ↑↓ ↑	— ↑↓ ↑↓ ↑↓ ↑↓	↑ ↑↓ ↑↓ ↑↓ ↑↓
Unpaired electrons	1	0	1	2	1	0	1	0	1

↑ energy E increases

WEAK SQUARE PLANAR

Number of d electrons	d^1	d^2	d^3	d^4	d^5	d^6	d^7	d^8	d^9
d electrons distributions $\overline{d_{x^2-y^2}}$ $\overline{d_{xy}}$ $\overline{d_{xz}}\,\overline{d_{yz}}$ $\overline{d_{z^2}}$	— — ↑ —	— ↑ ↑ —	— ↑ ↑ ↑	↑ ↑ ↑ ↑	↑ ↑ ↑↑ ↑	↑ ↑ ↑↓ ↑	↑ ↑ ↑↓ ↑↓	↑ ↑ ↑↓ ↑↓ ↑↓	↑ ↑↓ ↑↓ ↑↓ ↑↓
Unpaired electrons	1	2	3	4	5	4	3	2	1

↑ energy E increases

The interpretations given in Table 3.7 utilize the different orders of energy sequence that prevail with respect to the splitting of orbitals in the respective fields. (These have already been provided.)

In Table 3.7 the student will recognize the data appropriate to the d^8 nickel species; for example, the equal strengths of the strong and weak tetrahedral crystal fields of the $Ni(H_2O)_4^{2+}$ complex — in either instance, two unpaired electrons, paramagnetic. Likewise to be recognized is the strong square planar crystal field of the $Ni(CN)_4^{2-}$ complex — zero unpaired electrons, diamagnetic. The purely academic conjecture that the $Ni(H_2O)_4^{2+}$ complex might be predicted in a weak square planar field, rather than in a correct tetrahedral orientation, is amply resolved by considering the greater stability of the tetrahedral structures and the lack, in mixed-ligand derivatives thereof, of stereochemical *cis-trans* isomerism, as previously described.

The contrasts in magnetic criteria between the tetrahedral and the square planar coordinations of nickel complexes serves readily to differentiate one type of structure from the other. With paramagnetic compensations completely absent in square planar nickel complexes, these prove experimentally diamagnetic.

CF? VB? A CRITIQUE

In evaluating the respective serviceability of the two theories — electrostatic crystal field and valence-bond — in qualitatively interpreting and plausibly predicting bonding in transitional metallic complexes, it may be validly observed here that although the points of conflict between these concepts are relatively few, they are, nonetheless, relatively sharp in some instances.

As has already been noted, the valence bond theory is ineffective in explaining the structures of many copper(II) complexes in a fashion consistent with the indisputable experimental evidence of their octahedral distortions, rather than of square planar arrangements. Moreover, evidence seems to be wanting that the energy factors involved tend toward, or otherwise make feasible, the occupation of outer d orbitals by electrons when inner d orbitals (below the principal valence shell of a metal) have yet to be filled. The CF theory avoids these presumptions.

Moreover, the valence bond theory does not interpret the excited states of electrons nearly so well as it does those of the ground states. Although adaptive improvements in valence-bonding in this area of applicability are by no means improbable, the most advantageous and convenient way to depict qualitatively the d-orbital involvements of transition chemistry is by use of the electrostatic crystal field. The valuable qualitative applicability of the valence bond theory to nontransitional chemistry, particularly in the field of organic chemistry, would recommend thorough familiarity with it.

As has already been expressed elsewhere in our development of the theory of bonding, neither the electrostatic crystal field nor the valence bond concept provides the quantitative applicability or validity of molecular orbitals, although fair approximations of a quantitative character are obtainable with the electrostatic crystal fields when they are modified to include certain concessions to molecular orbital theory. These constitute the interpretative domain of the ligand field theory.

COLORS OF TRANSITION COMPLEXES

Despite the independability of any confident prediction of coloration, in view of the many variabilities involved, this topic deserves some consideration here. The colors of transition complexes, both neutral and ionic, may with a prudent exercise of logic, be attributed to the "d-to-d" electron transfers. By this we mean the movements of unpaired electrons or of electron-pairs from d orbitals of lower energy to available d orbitals of higher energy that lack their maximum individual complements of two electrons. As the primary source of energy absorptions that make possible the d-to-d excitations of paired and unpaired electrons is radiant energy in the form of light, we may reasonably deduce that:

1. Colorless or white transition complexes result when electrons are completely absent from d orbitals.

2. Colorless or white transition complexes result when absorbed energy is exclusively in the ultraviolet or the infrared region of radiation.

3. Transition complexes appear black when all wave lengths of visible light are totally absorbed.

4. When all but one of the colors of the visible spectrum are absorbed, the particular color that is transmitted to the human eye is the one color that was not absorbed.

5. When just one color of visible light is absorbed, the complex acquires a visible color complementary to the specific color absorbed.

The colors of complexes frequently provide highly informative corroboration of the anticipated strength of crystal fields and of the sizes and electron densities of the ligands therein. The distributions of d-electrons of the central transition-metallic ion or atom may also frequently be plausibly interpreted from colors. On the other hand, the complications of predicting color become fairly profound when more than one color of the visible spectrum is absorbed. What should be emphasized in any event, is that both complexes and uncomplexed species may be colored despite the absence of color in the individual atoms or ions that constitute the over-all species. This may occur even without d-to-d electron transfers, but is interpretable

COMPLEXING TRANSITIONAL METALS

in terms of the deformation of ions or atoms (*polarization*) that results from the opposing effects of electrostatic attraction and repulsion. The attraction is that of the positively charged nucleus of the central metallic atom or ion for the electron clouds of the anionic or molecular ligand; the mutual repulsion is between the nucleus of the central atom or ion and the individual nuclei of the ligands.

The foregoing ion-polarization effects have been formulated as *Fajans' Rules,* wherein a greater tendency toward covalency in bonding of anion and cation is predictable with:

1. *increased nuclear charge of the bonding opposite ion.* Thus, the more highly charged the cation, the greater is the polarization induced in the anion; and vice-versa. It follows, consequently, that the greater the charge upon the anion the more easily it becomes polarized by the cation.

2. *diminished size of the cation* and *increased size of the anion.* Thus, the smaller the cation the greater is the density of its positive charge and, consequently, the greater its influence to distort or polarize the anion. From this it also follows that the larger the size of a bonding anion, the more easily it becomes polarized. In general, anions may be expected to polarize more readily than cations, because anions normally offer greater numbers of electrons.

Considerations of color may be cautiously interpreted in terms of the following relationships between the component colors of the visible spectrum. The limits of radiation for perceptible visual color are between circa 7000 angstrom units (red region, the lower-energy radiation) to circa 4500 angstrom units (violet region, the higher-energy radiation):

COLOR						
absorbed	red	orange	yellow	green	blue	violet
seen	green	blue	violet	red	orange	yellow

EXERCISES

(See Note for Exercises of Chapter 1.)

1. The *radius ratio* of a complex (radius of the cation divided by the radius of the anion) permits an approximate determination of the maximum number of ligand anions or ligand molecules with which a given cation coordinates. The inherent

COMPLEXING TRANSITIONAL METALS

assumption of the rule is, that all the ligands are perfect spheres and that all are invariably in mutual contact with the central cation. This condition is hardly realistic; hence, its validity is only approximate. (The same rule may be applied, with similar prudent caution with respect to results, to the ion-aggregates that constitute the crystals of ionic solids.) In accordance with this formulation, the number of ligands that can be accommodated in a stable arrangement depends upon the sizes of the ligands and of the central cation — because ligancy is clearly increased by increase in cation radius and by decrease in anion radius. The value of the radius ratio R ($= r_+/r_-$, wherein the subscripts $+$ and $-$ conveniently denote cation and anion, respectively) determine the *limiting* structures of stable aggregates — the 8-coordination, characteristic of cubic form; the 6-coordination, characteristic of octahedral form; or the 4-coordination, characteristic of tetrahedral form; etc.

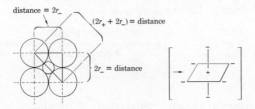

FIG. 3.8   Radial dimensions in evaluating geometric orientation.

We make an illustrative analysis of the usefulness of radial dimensions in the determination of geometric orientations by considering a regular octahedron. For this, a top-view cross-section may be visualized as follows:

As the square of the hypotenuse of the right triangle (bold outline) is equal to the sum of the squares of both arms, we obtain

$$(2r_+ + 2r_-)^2 = (2r_-)^2 + (2r_-)^2$$

which rearranges sequentially to the form of the quadratic equation,

$$4r_+^2 + 8r_+r_- + 4r_-^2 = 4r_-^2 + 4r_-^2$$

$$4r_+^2 + 8r_+r_- - 4r_-^2 = 0.$$

As we are interested in the ratio r_+/r_-, we divide this latter equation by $4r_-^2$, yielding

$$\left(\frac{r_+}{r_-}\right)^2 + 2\left(\frac{r_+}{r_-}\right) - 1 = 0.$$

This can be solved by the quadratic formula

$$x = \frac{-b \pm \sqrt{b^2 - 4ac}}{2a}$$

to give the roots

$$\left(\frac{r_+}{r_-}\right) = \frac{-2 \pm \sqrt{8}}{2} = \frac{-2 \pm 2.828}{2} = -2.414 \text{ and } + 0.414.$$

Therefore, $\frac{r_+}{r_-} = 0.414$ proves acceptable.

118   COMPLEXING TRANSITIONAL METALS

We can conclude that when $R = r_+/r_- = 0.414$, the lower limit of stability for 6-coordination symmetry has been reached; and that a value for R of less than 0.414 will most likely induce orientation of the ion-aggregates to a structure of diminished coordination number (e.g., 4-coordination tetrahedral). In this, the overcrowding of ligands, with their concomitant excessive repulsions due to "aggravated" overlapping of electron clouds, is effectively relieved.

Calculate the lower-limit value of the radius ratio R that is normally compatible with the stability of a cubic structure wherein the cation is "body-centered"; that is, the cation is in the center of the cube.

2. Given the following calculated limits of structural stability and ligand coordination as defined by the radius ratio $R = r_+/r_-$,

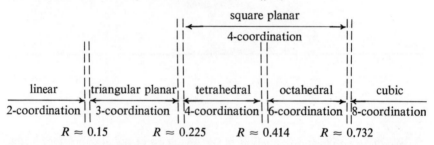

Interpret or justify therefrom the respective situations dictated for the crystal structures of each of the following ion-aggregates. Utilize for the purpose the ionic radii provided in the Periodic Table (Appendix D):
 (a) the most likely stable structure of CsBr.
 (b) the observed 4-coordination tetrahedral structure of BeO.
 (c) the observed 6-coordination octahedral structure of LiCl.

3. The $Co(NH_3)_6^{3+}$ ion is a low-spin complex; the CoF_6^{3-} ion is a high-spin complex. The former is diamagnetic; the latter is paramagnetic. Interpret this in terms of the relative extent of the splitting and the electron-occupations of the pertinent orbitals.

4. The responses of $Cr(H_2O)_6^{3+}$, $V(H_2O)_6^{3+}$, $Ti(H_2O)_6^{3+}$ ions to magnetic fields reveal the presence of different numbers of unpaired electrons.
 (a) give the electron compositions of the t_{2g} and e_g levels of each.
 (b) of the transition metals appearing in this exercise, which would offer simple cation(s) of smallest possible nuclear charge for central orientation that would prove, theoretically at least, susceptible to ligand-splitting of orbitals into strong (low-spin) and weak (high-spin) crystal fields? Justify your answers by giving all pertinent energy-level configurations, as well as the numbers of unpaired electrons.

COMPLEXING TRANSITIONAL METALS

5. *Carbonyls* of transition metals are formed, under suitable conditions, by direct coordination of ligand molecules of carbon monoxide ($\overset{..}{\underset{..}{:}}C\equiv O\overset{..}{\underset{..}{:}}$) with the free central metallic atom whose valence electrons are fully intact. These compounds are, in effect, neutral complexes. Although they are, generally, of simple monomeric formulation — as represented by $[M(CO)_n]^0$ — dimeric and other polymerized forms are known. As the central atom remains in an oxidation state of zero, some imagination is invited to reconcile their electronic configurations with their established molecular formulas.

A plausible approach to complexing, which appears to fit carbonyls of all transition metals is a conjectured shifting of electrons from the $4s$ orbital of the metal (assuming a carbonyl of the fourth transition series of the periodic classification) to a $3d$ orbital thereof. This may be accompanied, by a pairing of hitherto unpaired electrons.

Atoms of Ni^0, Fe^0, and Cr^0 all coordinate themselves fully with respect to CO-ligancy and, when a single electron-pair of each CO molecule being utilized is counted with the pertinent metallic atom, they acquire configurations that are isoelectronic with the structure of the inert gas, krypton, $Kr^0 = [Ar]\ 3d^{10}\ 4s^2\ 4p^6$. Utilize valence-bonding concepts to evaluate the most likely molecular formula for the neutral complex in each instance, as well as the arrangement and type of orbital hybridization to be assigned to it.

6. In each instance of carbonyl formation in Exercise 5, isoelectronic inert gas structures were attainable by the central metal because each supplied an even number of core-external electrons; that is, they all have even atomic numbers. Yet, carbonyls of metals of odd atomic numbers have been prepared, among them those of cobalt ($Z = 27$), leading to the structure of a metal that is isoelectronic with krypton ($Kr^0 = [Ar]\ 3d^{10}\ 4s^2\ 4p^6$); and of rhenium ($Z = 75$), leading to the structures of a metal that is isoelectronic with radon, $Rn^0 = [Xe]\ 4f^{14}\ 5d^{10}\ 6s^2\ 6p^6$.

(a) Utilize the concept of polymerization (the coupling of smaller units to form larger units) to depict by valence-bonding the simplest possible formula assignments to (1) cobalt carbonyl and (2) rhenium carbonyl.

(b) What would be the plausibly predictable formula for the carbonyl of manganese ($Z = 25$)?

7. The carbonyl $V(CO)_6$, unlike $Cr(CO)_6$, does not attain the structure of a central metal isoelectronic with krypton. Moreover, $V(CO)_6$ proves an exception to the experimental observations that virtually all carbonyls are diamagnetic. Account for this in terms of valence-bonding electronic distributions.

8. Assuming that light of only one specific wavelength is absorbed by a coordination compound, utilize the following relationships of the visible spectrum

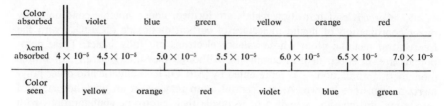

to determine

(a) the color of the complex as observed upon its absorption of 42.9 kilocalories for a mole of light radiation energy.

(b) the color of the visible light radiation absorbed by the complex if it is determined that the frequency of radiation corresponding to the observed color of the complex is 5.30×10^{14} sec^{-1}.

9. A certain dissolved complex compound having ligand-NH_3 molecules — all of which are in the coordination sphere — readily evolves NH_3 gas upon being heated; yet, the same complex does not readily permit neutralization of its NH_3 by strong acid. Is it correct to conclude that the compound's stability at high temperature is necessarily greater than its stability at diminished pH? Explain your answer.

10. The coordination compound $Ag_2[HgI_4]$, yellow in color, is obtained by warming a structurally different red form of identical molecular weight and atomic constitution. Suggest a ligand-distribution isomer for the red form, given that it, like its yellow-colored analog, is a 4-coordination iodo-complex.

11. *Optical* isomerism represents a separate category of stereoisomerism. While geometric isomers have distinctly different physical and chemical properties, optical isomers differ significantly only in the degree of their contrasting lability in responding to identical chemical changes, and in the opposing directions in which they rotate the plane of polarized light — that is, light vibrating in the single axial plane of a substance acting as a filter of multiwave white light. Optical isomers, always formed when there is no plane of molecular symmetry, may be likened to mirror images (such as the mutual relationships of a right hand and a left hand).

For the Co(III) complex $[Co(en)_2Br_2]^+$, wherein the abbreviation *en* stands for the ethylenediamine molecule — the bidentate ligand $NH_2CH_2CH_2NH_2$ — depict diagrammatically all the sterioisomeric forms possible. Indicate unambiguously the geometric *cis-trans-*, as well as the optical isomers.

12. *Ionization isomerism* interprets the differences in coordination-sphere make-up which account for the differences in chemical identities of free and un-

complexed ions external to the respective coordination spheres of mutually isomeric forms of a compound.

A solution of [Co(NH$_3$)$_4$Cl$_2$]I, when treated with AgNO$_3$, yields a yellow precipitate of AgI. What must be the formula of the coordination isomer of the dissolved compound that yields a white precipitate of AgCl. Assume for both isomers a tightly bonded coordination sphere.

13. *Coordination isomerism* interprets the mutual exchange of ligands between the coordination spheres of a complex cation and a complex anion of the compound.

Depict all the theoretically possible isomers of [Cu(NH$_3$)$_4$]$^{2+}$[PtCl$_4$]$^{2-}$, which would continue to provide simultaneously both a complex cation and a complex anion.

14. *Mode-Of-Attachment* isomerism interprets how a ligand that is itself isomeric links itself to the central metal. A typical example of this type of isomerism is that of nitrito- and nitro-compounds. The nitro compound manifests attachment of the ligand by its nitrogen atom; thus

$$\left[-N \lessgtr {}^O_O \right]^-.$$

The nitrito compound, in contrast, manifests attachment of the ligand by one of the oxygen atoms; thus [—O—N=O]$^-$.

Following this principle of *mode-of-attachment* isomerism, depict the substitution of a single ligand-NH$_3$ molecule of the coordination sphere [Co(NH$_3$)$_5$]$^{3+}$ by an NO$_2^-$ group, to yield two chemically different coordinated isomers.

15. *Multiform-Empiric* isomerism interprets the relationships that coordinated complexes bear to a common empirical formula. Thus, all the following complexes of element palladium — [Pd(NH$_3$)$_2$Cl$_2$]^0, [Pd(NH$_3$)$_4$]$^{2+}$[PdCl$_4$]$^{2-}$, and [Pd(NH$_3$)$_4$]$^{2+}$-([Pd(NH$_3$)Cl$_3$]$^-$)$_2$ have the identical empirical formula, [Pd(NH$_3$)$_2$Cl$_2$]$_n$, wherein the value of n is, respectively, 1, 2, and 3 for the given complexes. All similarity ends here, however, inasmuch as the structures of these complexes are all very different. These multiform-empiric coordination spheres may be transposed to exhibit compound anion-cation coordination isomerism, as well.

Depict all theoretically possible combinations of complex anion with complex cation for values of (a) $n = 2$ and (b) $n = 3$.

16. *Bridge isomerism* interprets the linkage of two complexed species by ligands that are common to the coordination spheres of both, and which, consequently, must be mutually shared. Illustratively, the existence of the ion, [Fe$_2$(OH)$_2$-

$(H_2O)_8]^{4+}$, is well established, and it is theoretically conjectured as a dual octahedral structure of 6-coordination with respect to each Fe^{3+} ion. In it the two OH^- ions act as bridges to connect the respective octahedral components of the over-all structure; thus

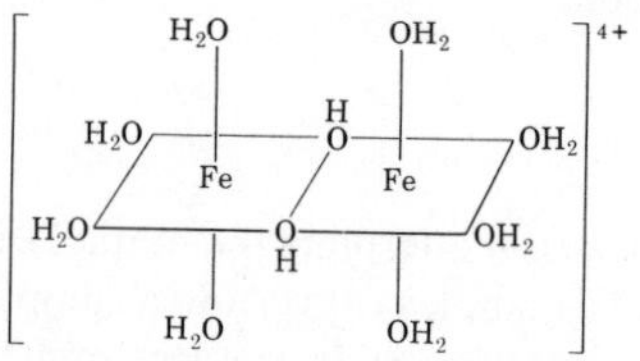

$$\left[\begin{array}{c} H_2O \quad\quad OH_2 \\ H_2O\diagdown \begin{array}{c} H \\ O \end{array} \diagup OH_2 \\ \quad\quad Fe \quad\quad Fe \\ H_2O\diagup \begin{array}{c} O \\ H \end{array} \diagdown OH_2 \\ H_2O \quad\quad OH_2 \end{array}\right]^{4+}$$

FIG. 3.9   Hydroxy-bridged structure (octahedral).

Assuming the possibility of substituting additional hydroxyl ions for aquo groups to yield the complex, $[Fe_2(OH)_4(H_2O)_6]^{2+}$, without affecting octahedral arrangements or the ultimate duality of the identical hydroxy-bridged components of the over-all structure, supply diagrams of geometric configurations of three possible isomers of the complex.

17. In the interhalogen ion, BrF_4^-, the less electronegative atom expands its octet to act as the central coordinator of the over-all structure.
   (a) How many total bonded plus unbonded electron-pairs surround the bromine atom?
   (b) Provide a plausible three-dimensional structure for the ion in (a) clearly showing the orientations of bonded and unbonded electron-pairs. (Designate the former by a pair of *dots*, and the latter by a pair of *crosses*.)
   (c) Designate the descriptive classification of structural arrangement conforming (1) to the bonding pairs alone; and (2) to all electron-pairs without differentiation.

18. In the *aufbau* (buildup) process of constructing the electronic configurations of the neutral gaseous transition *atoms* of the fourth periodic series, the energetics of filling have required the $4s$ orbital to be satisfied before the $3d$ orbital. Yet the electronic configurations of the simple gaseous divalent *ions* of these same atoms reveal that the electrons that have been lost in forming them have come from the $4s$ orbital instead.
   How is this apparent contradiction to be reconciled?

19. The cobalt(II) cyanide complex, $[Co(CN)_6]^{4-}$, is a d^2sp^3 orbital hybrid that manifests a paramagnetic response equivalent to one unpaired electron. Using

arrows to represent the central atom's electrons and *crosses* to denote ligand electrons, provide plausible electronic distributions within the complex in terms of
  (a) its valence-bonding.
  (b) its electrostatic crystal field.

20. In terms of valence-bonding, plausibly interpret the following reaction, which occurs when aqueous solutions of cobaltous ion, Co^{2+}, are treated with concentrated HCl or other sources of highly concentrated chloride ion. (Use *arrows* for the cobalt's electrons and *crosses* for those of the ligands.)

$$Co(H_2O)_6^{2+} + 4Cl^- \rightarrow CoCl_4^{2-} + 6H_2O$$
octahedral, pink   tetrahedral, blue

21. Provide a chemical formula for each of the following ions or compounds of the complexed type which, in conformity with established nomenclature, delineates unambiguously the constitution of the pertinent coordination sphere(s):
  (a) tetraiododiammineplatinum(IV).
  (b) dibromotetrammineplatinum(IV) bromide.
  (c) potassium tetrachloroaquoamminecobalt(II).
  (d) potassium dichlorodiaquodiamminecobalt(III).
  (e) dichlorobis(ethylenediamine)nickel(IV) ion.
  (f) pentafluoroaquonickelate (IV) ion.
  (g) tetraiodo(ethylenediamine)chromate(III) ion.
  (h) oxalatotetraquoiron(III) sulfate.
  (i) iron(II) hexafluoroferrate(III).
  (j) hexamminecobalt(III) hexachlorochromate(III).
  (k) diamminesilver(I) tetrachloroplatinate(II).

CHAPTER

FOUR

ASPECTS OF MO THEORY

THE PREVIOUS DESCRIPTIONS OF CHEMICAL BONDING HAVE NECESSARILY followed the path of least resistance; that is, the easiest routes represented by the atomic orbitals of the VB and CF theories. These concepts have, indeed, supplied the chemist with plentiful and highly rewarding information concerning the constructions that can plausibly be placed upon the nature and chemical behavior of matter. Nonetheless, the virtue of simplicity must not be allowed to lure us away from the realization that all nuclei and all electrons in a chemical species belong to the chemical unit as a whole in molecular orbitals. Hence, they contribute to the physical and chemical attributes of the whole unit and not merely to any specific constituent atom alone.

The purpose of this chapter, then, is to explain the exceptions to the rule, with which atomic orbital theories abound, and to familiarize the student with the language of molecular orbital concepts.

THE ENERGY AND STABILITY OF MOLECULAR ORBITALS

We initiate our discussions with considerations leading to an Energy-Stability diagram. This permits us to clarify the electronic orientations of any diatomic species — whether a neutral molecule or its net-electrically-charged analog called a "molecule-ion," in a way very similar to that used to deduce the electronic configurations of atoms and their pertinent ionic analogs by signations as s, p, d, and f. In fact, there is a certain superficial parallelism in the systematic filling-in of molecular orbitals that is reminiscent of the VB and electrostatic CF procedures involving the d orbitals in the complexing of transition-metal ions.

Definitions of MO (molecular orbital) terms and symbols to be utilized in the diagram for energy-stability, shown in Figure 4.1 are as follows:

1. σ (sigma), π (pi), and δ (delta) define the various types of molecular orbitals to be encountered. These are used instead of the parallel s, p, and d designations of atomic orbitals.

2. Any molecular orbital bearing an asterisk* as a superscript is termed an *antibonding orbital;* any molecular orbital not so modified is either a *bonding orbital* or a *nonbonding orbital*. The precise distinctions between these two latter terms are made clear as we proceed.

3. The subscript appended to the designation of a molecular orbital indicates the specific atomic orbital that contributed to its formation; illustratively,

σ_{1s} signifies a bonding molecular orbital to whose formation a $1s$ atomic orbital has contributed; and

π_{2p}^* signifies an antibonding molecular orbital to whose formation a $2p$ atomic orbital has contributed.

4. Each heavy line (dash) in the energy level diagram represents an orbital which, as usual, can accommodate one, or, at most, two electrons (of opposing spins). It is to be stressed that the space intervals between the orbital levels in the MO diagrams provided, by no means represent the actual experimental relationships of difference in energy; rather, they arbitrarily denote the relative sequence of such energy levels. Indeed, the orbital energies of the π_{2p} and σ_{2p} may be so very close to one another that for some diatomic molecules and molecule-ions the specific orders of their electron fill-in are the reverse of those established for others. Two lines on the same level denote two *degenerate* orbitals of equivalent energy; that is, the two π_{2p} orbitals, or the two π_{2p}^* orbitals of the MO diagram shown in Figure 4.1.

5. The Energy-Stability diagram provides two sigma orbitals of different energy and stability for each *two* atomic orbitals that mutually coalesce in molecular orbital formation. One of the sigma MO's (the lower energy orbital) is always a bonding orbital, and the other (the higher energy orbital) is always an antibonding orbital, as follows:

atom A		atom A'		molecule AA'
$2s$	$+$	$2s$	$\rightarrow$	σ_{2s} σ_{2s}^*
1 atomic orbital		1 atomic orbital		2 molecular orbitals

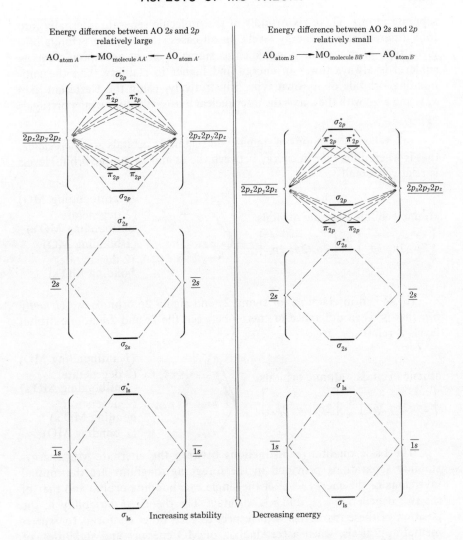

FIG. 4.1 Ground state energy-stability MO diagram for homonuclear diatomic species.

The interaction of two atomic orbitals to form a molecule must always produce two molecular orbitals common to that molecule. Inasmuch as each single atom in its isolated condition has three degenerate p orbitals, six molecular orbitals are formed by the union of the six-in-total separate but energetically equivalent p atomic orbitals supplied by two otherwise separate atoms. These six orbitals of the molecule separate internally into four different sets of energy levels. In all cases, the bonding orbitals of a given type of molecular bond, such as sigma or pi, in any specific energy sublevel is always lower in energy and higher in stability than the antibonding orbitals of its own type. Illustratively, using the Cartesian axes x, y, and z — with the x axis the internuclear axis — the following portrayals apply:

(a) When interactions of sigma $2s$ and sigma $2p$ orbitals are *appreciable;* that is when difference in energy between the $2s$ and $2p$ atomic orbital levels is relatively small.

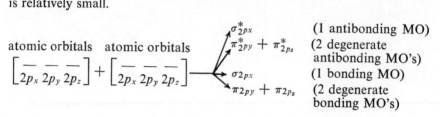

(b) When interactions of sigma $2s$ and sigma $2p$ orbitals are *virtually nil;* that is when difference in energy between the $2s$ and $2p$ atomic orbital levels is relatively large.

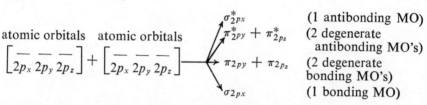

The basic qualitative distinctions between the alternate MO energy-stability separations provided in the foregoing diagrams are the mutual inversions of the energy level of the single σ_{2px} bonding orbital and that of the two degenerate π_{2py} and π_{2pz} orbitals. The theorized variability in the position of these two levels with respect to each other conforms to experimental appraisals, when ascertainable, of MO energies and stabilities, of numbers of bonds shared between atoms, and especially of paramagnetic properties of the pertinent species.

We should hasten to redeem somewhat the crudeness of the pictorial separations of energy levels that we have made by stressing, again, that it is the *relative sequence*, and not the *quantitative aspects*, of energy that concern us at the moment. The energy of the molecular bonding orbital is always lower and the stability is greater than that, respectively, of the atomic orbitals whence they formed; and the energy of the molecular antibonding orbital is always greater and its stability smaller than the corresponding attributes of the atomic orbitals of its construction. Moreover, it may be plausibly concluded that the extent to which the energy and stability of the atomic orbitals are, respectively, altered upon coalescence into the bonding molecular orbital is approximately equal to the extents of alterations likewise manifested within the antibonding orbital of the molecule. To be precise, the antibonding orbital is just a bit more positive in the degree of its energy separation from the accumulated energy of the contributory atomic orbitals than the analogous separation of the bonding molecular orbital is negative. This can be seen in the following disgram:

$$\text{greater energy } E \quad \underline{MO^*} \text{ (antibonding)}$$

energy AO ———— $\Delta E_1 + E$ ———— AO energy ; $(\Delta E_{AO \rightarrow MO^*} \approx \Delta E_{AO \rightarrow MO})$

$$\Delta E_1 - E$$

(bonding) $\overline{MO}$ greater stability

BONDING AND ANTIBONDING MO'S

Differentiation between "bonding" and "antibonding" orbitals must be justified, particularly in view of the connotations of "attraction" and "repulsion" that, respectively, accompany the two terms, but which, in actuality, express merely relativity or degree, not absolute or intrinsic nature. For, not only do all electrons in all orbitals — bonding as well as antibonding — exert mutual repulsion but, indeed, antibonding electrons may themselves be utilized in bonding, although by no means automatically so. When an electron is in an excited state its mere presence in an internuclear region (and presumed, therefore, to be shared) does not necessarily lead to the formation of a bond between the two atoms.

It is upon this conjecture that a logical differentiation between the bonding and antibonding MO is best made. An electron in an antibonding orbital does not contribute to molecular bond formation because the density of its cloud distribution between two atomic nuclei is virtually nil. Consequently, the collective total energy of the system of two separate and independent atomic orbitals cannot be significantly decreased. No bond can form unless the total energy of the final system is less than the total energy of the initial system. On the other hand, an electron in a bonding orbital does contribute to bond formation because the distribution of its cloud density in the internuclear region does diminish the total collective energy of the initial system of originally separate component atoms.

Another way of looking at this is that the presence of an adequate concentration of electrons between two nuclei effectively screens out the mutual repelling forces that unshielded nuclei would otherwise exert upon one another and, thus, permits their mutual association, or bonding, within the same particulate unit. This is the characteristic assigned to the lower-energy MO — the bonding MO. When, on the other hand, the predominant electron concentration remains in, or is diverted to, the peripheral regions of the atomic nuclei, there is virtually no shielding between them and the consequent repulsion is strong enough to prevent the overlapping required for covalent bonding. This is the characteristic assigned to the higher-energy, or antibonding, MO.

The mode of formation of two MO's — one bonding, the other antibonding — from two atomic orbitals may be conveniently represented by visualizing the atomic orbitals as purely mathematically differentiated into positive ($+$) and negative ($-$) regions or lobes with respect to the displacement of their electron clouds relative to a common reference — the nodal axis of zero electron displacement. This may be likened to plucking in the middle a taut string between its stationary supports; the uncomplicated up-and-down vibration of the resultant wave yields an upper region of movement (mathematically plus) and a lower region of movement (mathematically minus). Both the crest (upper curve) and trough (lower curve) of the wave are symmetrically oriented around the nodal axis that contains the node, itself. These considerations are represented in Figure 4.2.

FIG. 4.2 Nodal-axis description of atomic orbitals.

A molecular orbital is presumed to be constructed in accord with the principle of *Linear Combination of Atomic Orbitals* (LCAO). In this, the wavefunction psi (ψ), which represents the orbital of a single electron of one atom, is algebraically combined with the wavefunction corresponding to the orbital of a single electron of a second atom to yield the wavefunction of a common molecular orbital that contains both electrons. This may be shown as

$$\psi_{MO} = c'\psi'_{AO} + c''\psi''_{AO}$$

wherein c' and c'' are constants. These constants numerically evaluate the relative contribution of each of the respective atomic wavefunctions (ψ'_{AO} and ψ''_{AO}) to the molecular wavefunction (ψ_{MO}). They are, consequently, factors that depend upon the nature of the respective atomic orbitals that are being incorporated by the LCAO procedure. The two constants are exactly equal to each other only in diatomic homonuclear molecules. According to the assignments made to "plus" and "minus" wave displacements which are algebraically relative to the reference made for zero-displacement, it can be demonstrated mathematically that ψ may be both plus and minus. It then follows that the formulation for the generalized diatomic AB molecule from its initially separate and independent atoms A and B must actually be expressed by two separate equations for ψ_{MO}:

(i) $\quad \psi_{AB} = c'\psi_A + c''\psi_B$

and

(ii) $\quad \psi^*_{AB} = c'\psi_A - c''\psi_B.$

The first expression, an addition of atomic orbital wavefunctions, always signifies an over-all net of attraction between atoms; hence, ψ_{AB} symbolizes a bonding MO and the electrons therein are termed *bonding electrons*. The second expression, a subtraction of atomic orbital wave functions, always signifies an over-all net of repulsion between atoms; hence, ψ^*_{AB} symbolizes an antibonding MO and the electrons therein termed *antibonding electrons*. Clearly, a stable molecule cannot be constructed unless the numbers of bonding electrons exceed the numbers of antibonding electrons; that is, the *bond order* must be an algebraically positive value.

How do we interpret the alterations undergone by the orbital cloud densities of the originally separate atoms in the process of molecule formation? How, that is, do we justify our visualizations of molecular orbitals as either bonding or antibonding? Manifestly, for the formation of a bonding MO, interaction of the two AO's must produce a concentration of electron density between the nuclei of the pertinent atoms sufficient to screen out, effectively, the repulsion each nucleus would otherwise exert upon the other. To achieve this condition there must be considerable overlap of the electron clouds of the atoms. This means that the charge on electrons must be *delocalized* from the individual atoms and shifted predominantly to the internuclear region.

The degree of delocalization of charge determines the degree of stability of the bond. The greater the delocalization of charge from the atoms A and B, the more stable is the molecule AB, and the smaller, consequently, the energy of the bond. The tendency of the constituent atoms thereof to revert to their initial independent states, or to become involved in intermolecular reactions that would sever their mutual molecular associations, is also less.

On the other hand, for the formation of an antibonding orbital no significant concentration of electron charge may be permitted in the internuclear region. Delocalization of the charges upon the atomic orbitals and their relocalizations within the internuclear region must than be so slight that there is virtually no shielding of nuclear repulsion. Consequently, the energy of antibonding MO's is always higher than that of their bonding analogs, and sufficiently higher to discourage the overlaps that must be reconciled with the formation of permanent stable bonds.

"IN PHASE" AND "OUT OF PHASE" VARIATIONS OF MO'S

The mathematical implications of the LCAO procedure in the formation of bonding MO's (lowest energy) and antibonding MO's (highest energy) may be projected pictorially to the alterations in the shapes of the atomic orbitals in the process of overlap or coalescence. These are best pursued in their illustrations by the alternations of combination referred to as "in phase" and "out of phase." These accord with the algebraic charge assignments already made with respect to AO electron wave displacements from the common reference node, as shown in Figure 4.3.

We now utilize the elements of Periodic Group 2 to fix or determine, in conformity with the diagram for MO energy-stability (Fig. 4.1) the manner of molecular-orbital filling that establishes for the species not only the theoretical justification for the number of bonds or electron-pairs assigned to sharing between the atoms, but also the reconciliation with experimental data that defines the precise number of electrons unpaired — hence, the degree of paramagnetism of the species. In establishing structure, then, the theory must conform to experimental fact. But first we must establish the "ground rules" of the systematic procedure to be employed. We need involve ourselves only with sigma- and pi-bonds to establish these rules adequately.

MO CONSTRUCTIONS AND BONDINGS

1. Electrons are accommodated within molecular orbitals in accordance with the restrictions imposed by the *Pauli Exclusion Principle* and by Hund's *Rule of Maximum Multiplicity*, both of which constitute the *aufbau*, or buildup, employed so plausibly in delineating the atomic-orbital filling. Hence, electrons enter a molecular orbital of a specific energy level one at a time before they can pair with opposing spins. This means that for the two degenerate π_{2p} orbitals, or for the two degenerate π_{2p}^* orbitals, the electrons must enter into each singly — and with parallel spins — before they can mate within a common orbital.

It is this distribution which, by separating energetically equivalent electrons as far apart as possible, minimizes their mutual repulsions. The *aufbau* principle means, too, that each molecular orbital of a given energy

ASPECTS OF MO THEORY

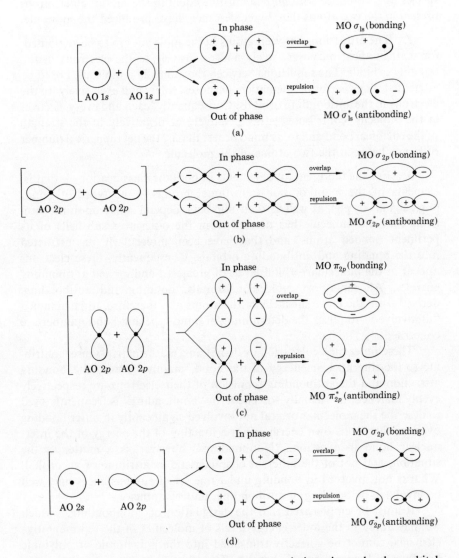

FIG. 4.3 "In phase" and "Out of phase" variations in molecular orbitals. (a) s-s overlap in sigma bonding. (b) p-p head-to-head overlap in sigma bonding. (c) p-p side-to-side overlap in pi-bonding. (d) s-p overlap in sigma bonding.

level receives its full complement of two electrons before an electron can be "invited" to an orbital of higher energy (smaller stability). The total number of electrons to be introduced into the collective molecular orbitals of the species in accordance with *aufbau* must be equal to the sum of all electrons present in the two atoms that, by coalescence, have produced the molecule.

2. The attractions between atoms in the molecule are to be attributed, for mathematical purposes, solely and exclusively to the electrons in the bonding orbitals. The repulsions between the atoms in a molecule are to be attributed, likewise for mathematical purposes, solely and exclusively, to the electrons in the antibonding orbitals. Consequently, each and every electron in the molecule contributes either positively or negatively to the strength of the chemical bond and to its bond order; that is, the net numerical number of bonds between the two atoms of the molecule.

3. For the representative elements of periodic classification — that is, elements of the main or non-transitional groups — the types of bonds (e.g., sigma- or pi-), as well as their numbers, depend solely upon the electrons that the molecule has derived from the outermost subshells of its pertinent bonded atoms, and that have been molecularly reconstructed into the bonding and antibonding orbitals. Consequently, these electrons appear in the molecular-orbital levels of greatest bonding and antibonding energies. All innermost molecular orbitals, bonding and antibonding, derived from the inner subshells of the atoms are, therefore, to be assumed "uninvolved" so far as the determining of bond type and bond number are concerned.

These inner molecular orbitals do not, in a mathematical sense, contribute to the internuclear energy of the bond, mainly because the bonding attraction and the antibonding repulsion of their electrons are respectively evenly matched, or virtually so. The term "nonbonding" is frequently used to describe the molecular orbital not involved significantly in outer bonding overlap; that is, its own energy is not a function of the energy of the internuclear bond in question. "Nonbonding," however, is a matter of the situation, and not of the inherent characteristic or attribute of an orbital. What is not involved in bonding under one set of circumstances may well be involved, by bonding or antibonding, under another.

It must be emphasized that the designation of "nonbonding," which we may apply to the inner energy levels of molecules of the representative elements, cannot be correctly translated into the polyatomic or polyionic bondings of the transitional elements. These latter characteristically involve in their bondings the belated utilization of orbitals and electrons of inner energy levels (see Chapter 3, dealing with transition complexes). This belated filling of innermost d and f atomic orbitals proved, in fact, the definitive characteristic of transition-metal ion bondings used to distinguish them from those of the representative elements.

Nonbonding orbitals of homonuclear diatomic molecules include, but do not wholly comprise, the closed-shell orbitals of the bonding atoms that comprise the molecule. Just as [He] is the closed K shell of atoms of the second periodic series, or [Ne] represents the closed K and L shells of the atoms of the third periodic series, or [Ar] the closed K, L, and M shells of the fourth periodic series of atoms, etc., so the closed molecular shells may be symbolized as KK, $KKLL$, $KKLLMM$, etc., representing the overlapped atomic cores, He–He, Ne–Ne, and Ar–Ar, respectively.

Thus, the molecular designation of the N_2 molecule,

$$\sigma_{1s}^2 \sigma_{1s}^{*2} \sigma_{2s}^2 \sigma_{2s}^{*2} \pi_{2p}^2 \pi_{2p}^2 \sigma_{2p}^2$$

wherein the superscripts denote the numbers of electrons in the respective orbitals, may be written just as revealingly as

$$KK \, \sigma_{2s}^2 \sigma_{2s}^{*2} (\pi_{2p}^2 \pi_{2p}^2 \sigma_{2p}^2).$$

In this latter designation, the parenthesized orbitals are really all that are explicitly required for calculating bond number, bond type, and paramagnetic response of the molecule. The σ_{2s}^2 and σ_{2s}^{*2} orbitals, like those of the KK molecular shell, are nonbonding here. In the N_2 molecule, therefore, we have just three orbitals that are not nonbonding, all other bonding and antibonding orbitals having virtually cancelled out their opposing influences; that is, the numbers of inner bonding electrons are equal to the numbers of inner antibonding electrons.

4. The bond order, or net number of bonds, is obtained by subtracting the total number of antibonding electrons from the total number of bonding electrons, and dividing the result by 2; that is

$$\frac{\text{net number of bonds}}{\text{(bond order)}} = \frac{\text{total bonding electrons} - \text{total antibonding electrons}}{2}.$$

The formula provided is equivalent to a statement that the total number of bonds formed is equal to the total number of electron pairs in the bonding MO's, minus the total number of electron-pairs in the antibonding MO's. The point to be stressed, however, is that the molecular-orbital concept makes no assumptions whatever of the pairing of electrons in the bondings of atoms.

The illustrative analyses of molecular orbital configurations, as given in Table 4.1, show that *single*-electron bonds or half-bonds may prevail in the molecular orbital concept. It is to be anticipated that a greater degree of stability normally favors the species possessed of a numerically larger bond order. In any event, a *positive* algebraic value thereof is required to support the conclusion that a species may exist at all. This is of importance in some situations in interpreting relative degrees of stability from calculated bond orders. Clearly, if an electron in an antibonding orbital exerts a slightly greater force for repulsion of the two pertinent atoms than does an electron in a bonding orbital analogously for their attraction, the bond orders of all species that contain antibonding electrons must, in actuality, be slightly less positive than those calculated without the benefit of this incremental difference. This consideration frequently permits a logical interpretation of differences in stability of two species that have identical calculated bond orders but differing numbers of antibonding electrons.

GROUND STATE MOLECULAR ORBITALS OF HOMONUCLEAR DIATOMIC SPECIES. ELEMENTS OF PERIODIC SERIES 1 AND 2

The progressive sequence of filling the molecular orbitals with electrons may be visually simplified as follows:

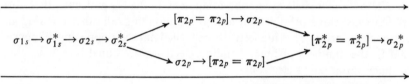

This arrangement assists the analyses shown in Table 4.1. In each instance, the superscript to the MO indicates the number of electrons that conform to the established normal ground rules for orbital filling (as far as is known experimentally). The parenthesized orbitals of the MO *configuration* denote those pertinent to the evaluations of the bond number, bond type, and paramagnetic response of the molecule.

The analyses and progressive supporting conclusions will be assisted by reading the columns in Table 4.1 systematically from left to right.

These few applications of the molecular orbital concept bring to the fore the lack of validity of some visualizations of atomic structure that the atomic orbital concept sought to establish. For one, the prediction of the molecular orbital concept that the O_2 molecule will prove paramagnetic to the extent of two unpaired electrons (which is completely borne out by experiment) reveals the *inaccuracy* of the electron-dot formulation of this molecule as

$$:\ddot{O}:\ :\ddot{O}:$$

Although this formula satisfies the octet rule that valence-bonding requires the installation of eight electrons around each of the oxygen atoms, (in due conformity with the calculated net double bond of the molecule), it clearly

Table 4.1

Species	Total electrons	Atomic components of formation	MO configuration	Bond order	Bond type	Unpaired electrons
H_2^+	1	$2(H^0 = 1s^1) - 1e^-$	σ_{1s}^1	$[(1-0)/2] = 0.5$	half ($1e^-$) sigma	1
H_2	2	$2(H^0 = 1s^1)$	σ_{1s}^2	$[(2-0)/2] = 1$	single sigma	0
H_2^-	3	$2(H^0 = 1s^1) + 1e^-$	$\sigma_{1s}^2 \sigma_{1s}^{*1}$	$[(2-1)/2] = 0.5$	half ($1e^-$) sigma	1
He_2	4	$2(He^0 = 1s^2)$	$\sigma_{1s}^2 \sigma_{1s}^{*2}$	$[(2-2)/2] = 0$	nonexistent	—
Li_2	6	$2(Li^0 = 1s^2 2s^1)$	$KK(\sigma_{2s}^2)$	$[(2-0)/2] = 1$	single sigma	0
Be_2	8	$2(Be^0 = 1s^2 2s^2)$	$KK(\sigma_{2s}^2 \sigma_{2s}^{*2})$	$[(2-2)/2] = 0$	nonexistent	—
B_2	10	$2(B^0 = 1s^2 2s^2 2p^1)$	$KK\sigma_{2s}^2 \sigma_{2s}^{*2}(\pi_{2p}^1 \pi_{2p}^1)$	$[(2-0)/2] = 1$	single ≈ one pi, *net*	2
C_2	12	$2(C^0 = 1s^2 2s^2 2p^2)$	$KK\sigma_{2s}^2 \sigma_{2s}^{*2}(\pi_{2p}^2 \pi_{2p}^2)$	$[(4-0)/2] = 2$	double: two pi	0
N_2	14	$2(N^0 = 1s^2 2s^2 2p^3)$	$KK\sigma_{2s}^2 \sigma_{2s}^{*2}(\pi_{2p}^2 \pi_{2p}^2 \sigma_{2p}^2)$	$[(6-0)/2] = 3$	triple: one sigma, two pi	0
O_2	16	$2(O^0 = 1s^2 2s^2 2p^4)$	$KK\sigma_{2s}^2 \sigma_{2s}^{*2} \begin{array}{c}(\pi_{2p}^2 \pi_{2p}^2 \sigma_{2p}^2 \pi_{2p}^{*1} \pi_{2p}^{*1}) \\ or \\ (\sigma_{2p}^2 \pi_{2p}^2 \pi_{2p}^2 \pi_{2p}^{*1} \pi_{2p}^{*1})\end{array}$	$[(6-2)/2] = 2$	double: one sigma, one pi	2
O_2^-	17	$2(O^0 = 1s^2 2s^2 2p^4) + 1e^-$	$KK\sigma_{2s}^2 \sigma_{2s}^{*2} \begin{array}{c}(\pi_{2p}^2 \pi_{2p}^2 \sigma_{2p}^2 \pi_{2p}^{*2} \pi_{2p}^{*1}) \\ or \\ (\sigma_{2p}^2 \pi_{2p}^2 \pi_{2p}^2 \pi_{2p}^{*2} \pi_{2p}^{*1})\end{array}$	$[(6-3)/2] = 1.5$	single sigma; half ($1e^-$) pi	1
O_2^{2-}	18	$2(O^0 = 1s^2 2s^2 2p^4) + 2e^-$	$KK\sigma_{2s}^2 \sigma_{2s}^{*2} \begin{array}{c}(\pi_{2p}^2 \pi_{2p}^2 \sigma_{2p}^2 \pi_{2p}^{*2} \pi_{2p}^{*2}) \\ or \\ (\sigma_{2p}^2 \pi_{2p}^2 \pi_{2p}^2 \pi_{2p}^{*2} \pi_{2p}^{*2})\end{array}$	$[(6-4)/2] = 1$	single sigma	0
F_2	18	$2(F^0 = 1s^2 2s^2 2p^5)$	$KK\sigma_{2s}^2 \sigma_{2s}^{*2} \begin{array}{c}(\pi_{2p}^2 \pi_{2p}^2 \sigma_{2p}^2 \pi_{2p}^{*2} \pi_{2p}^{*2}) \\ or \\ (\sigma_{2p}^2 \pi_{2p}^2 \pi_{2p}^2 \pi_{2p}^{*2} \pi_{2p}^{*2})\end{array}$	$[(6-4)/2] = 1$	single sigma	0
Ne_2	20	$2(Ne^0 = 1s^2 2s^2 2p^6)$	$KK\sigma_{2s}^2 \sigma_{2s}^{*2} \begin{array}{c}(\pi_{2p}^2 \pi_{2p}^2 \sigma_{2p}^2 \pi_{2p}^{*2} \pi_{2p}^{*2} \sigma_{2p}^{*2}) \\ or \\ (\sigma_{2p}^2 \pi_{2p}^2 \pi_{2p}^2 \pi_{2p}^{*2} \pi_{2p}^{*2} \sigma_{2p}^{*2})\end{array}$	$[(6-6)/2] = 0$	nonexistent	—

conflicts with the experimental observation that the molecule has a permanent magnetic moment equivalent to the spins of two unpaired electrons.

On the other hand, were we to picture the molecule as

$$:\ddot{O}:\ddot{O}:$$

the requirements of paramagnetism would be satisfied, but the octet rule would be violated because there are only seven electrons around each atom. The latter formulation, moreover, is inconsistent with the belief that there is more than just a single bond between the oxygen atoms, because more energy is required to break the molecular bond than can possibly be reconciled with the single shared pair of electrons. No traditional empirical electron-dot formula can be provided that is at once consistent with both the relatively high bond dissociation energy of the molecule and its unpaired electrons. And, as none is provided, the molecular orbital concept must stand merely on the calculated description of these properties.

Let us now apply the delineated principles of the MO concept to a simple heteronuclear diatomic species. A familiar example is the molecule NO. The electrons contributed by the pertinent atoms therein are N = [He]-$2s^2 2p^3$ and O = [He]$2s^2 2p^4$. We assume that overlapping interactions of σ_{2s} and σ_{2p} orbitals are sufficiently large to induced filling of π_{2p} before σ_{2p}. Our molecular configuration must conform to a total of eleven electrons, external, to the closed KK shell. Of these, four are nonbonding — those of σ_{2s}^2 and σ_{2s}^{*2} orbitals. The remaining seven are all pertinent to the computations of bond order and bond type. Hence, we may write for the molecular configuration the following:

$$NO = KK\ \sigma_{2s}^2\ \sigma_{2s}^{*2}\ (\pi_{2p}^2\ \pi_{2p}^2\ \sigma_{2p}^2\ \pi_{2p}^{*1}\ \pi_{2p}^{*0}).$$

Bond order computes to $[(6 - 1)/2] = 2.5$; and with one antibonding electron qualitatively negating one bonding electron, we conclude that we have here a molecule with bond construction of a single σ, a single π, and a half π (a one-electron bond). We might easily project this into a plausible electronic formulation, in electron-dot style, by assuming that we have three bonds — one σ and the other two π — that are collectively diminished in net energy by the presence of a single antibonding electron. With an asterisk to symbolize the antibonding electron and dots to designate all other electrons we may electronically represent the formula as

$$\left[:N\overset{*}{::}O:\right]$$

Electrons between the atoms are considered as bonding, and those outside each atom as nonbonding.

As observed, no differentiation of the nonbonding electrons with respect to their orbital character (σ_{2s} or σ_{2s}^*) has been attempted. Their peripheral distribution in equal numbers on each of the atoms is all that is necessary to our description. We must bear in mind, however, that the electrons described here as nonbonding are restricted to their involvements in the interatomic bondings of their own molecules. In no sense does the term preclude their being utilized in bondings with *other* molecules that, deficient in electrons, may "invite" electron-donors into chemical reaction.

Similarly the CO$^+$ molecule-ion may be described as constructed from C^0 = [He]$2s^22p^2$ and O^0 = [He]$2s^22p^4$, with subsequent removal of one electron. We must account for nine valence electrons; hence, we may write the molecular configuration as

$$CO^+ = KK\, \sigma_{2s}^2\, \sigma_{2s}^{*2}\, (\pi_{2p}^2\, \pi_{2p}^2\, \sigma_{2p}^1).$$

Bond order computes to $[(5 - 0)/2] = 2.5$. As there are no antibonding electrons in the $2p$ orbitals, any adherence to an electronic notation consistent with that just provided for the NO molecule must inevitably leave us with [:C :·: O:]$^+$. As observed, simple electronic designs such as these do not always fulfil the octet rule; nor are they necessary to our description of the molecule's properties.

The frustrations noted with respect to the electronic representation of the O$_2$ molecule might have been ameliorated by the same purely speculative bonding attachment (in degree) utilized in the NO molecule. Using the same type of symbolism, we might infer from the MO configuration previously tabulated for the O$_2$ molecule that, in effect, there are three bonds between the two oxygen atoms — one σ and two π. If these are collectively reduced to the evaluated two-bond (one σ, one π) net energy by the presence of two π antibonding electrons, the electronic formulation of the molecule might then be described by $\left[:O \overset{**}{::} O: \right]$

We have hardly scratched the surface of the MO concept. The involvements are numerous and often complicated. The ground states that constitute the relative orbital energies of molecules under ordinary environmental conditions must be evaluated in terms of frequently appreciable electronegativity differences of the bonding atoms. Their bonding electrons must be presumed to spend more time upon the more electronegative atom while the antibonding electrons spend more time upon the less electronegative. These considerations must receive more than casual attention in the higher polynumeric atomic bondings of organic molecules if the justifications and predictabilities of their chemical behavior are logically to be expressed.

MOLECULAR ORBITALS IN COMPLEX IONS

We concern ourselves now with the *modus operandi* of the MO theory in the chemical bondings that characterize the coordination spheres of transition-metal ions. It proves of interest, moreover, to compare the various pictures we develop of the MO ion-complexing process with those already viewed in the frames of their electrostatic crystal fields (as already amply described in Chapter 3). Actually, no new terms have to be introduced. Not only are the differentiations of nonbonding, antibonding, and bonding orbitals still maintained, as well as the assimilations of orbitals into sigma- and pi-bonds, but also the orderly order of electron-filling (*aufbau*) consistent with the familiar Pauli *Exclusion Principle* and Hund's Rule of *Maximum Multiplicity*. In the main, the contrasts between MO applications to complex ions and to the diatomic molecules previously discussed, are to be found in the sequence and numbers of the energy levels and the degrees of their *orbital degeneracies*.

The MO theory of transition-metal complexing starts with the mutual overlap of the electron clouds of the orbitals of the transition-metal ion and those of the ligands within its sphere. Inasmuch as a transition metal characteristically supplies nine atomic orbitals for accommodating electrons — five inner d orbitals (d_{xy}, d_{xz}, d_{yz}, d_{z^2}, $d_{x^2-y^2}$), one outer s orbital, and three outer p orbitals (p_x, p_y, p_z) — it is necessary to ascertain which of these actually lend themselves to bonding with the ligands. It must be presumed that their different orientation with respect to the Cartesian coordinates will not permit all five of the d orbitals of the transition-metal ion to overlap with equal effectiveness with the orbitals of the ligands. The d_{xy}, d_{xz}, and d_{yz} orbitals of the transition metal have their electron clouds oriented *between* the axes, while the ligands are approaching *along* the axes, in direct confrontation with the transitional d_{z^2} and $d_{x^2-y^2}$ electron clouds already lying along the axes. In their head-on encounters with ligands, the d_{z^2} and $d_{x^2-y^2}$ orbitals will be much more effectively overlapped.

Naturally, the same considerations of direct head-on meeting must prevail with the single s orbital of the transition metal (symmetrically concentric around the origin) and its three p orbitals, each of which is oriented along its respective Cartesian axis (x, or y, or z). Thus, only six of the nine available atomic valence orbitals of the transition metal are actually involved in the overlaps of effective MO bonding that are conducive to the formation of six molecular orbitals or bonds with six σ ligand valence orbitals. The other three d orbitals of the transition-metal ion — the d_{xy}, d_{xz}, and d_{yz} — are then necessarily to be regarded as virtually nonbonding, at least in octahedral complexes wherein the σ MO's alone, are presumed to describe satisfactorily the observed or measured characteristics of the complex.

It must not be concluded, however, that the d_{xy}, d_{xz}, and d_{yz} orbitals are completely immune from bonding involvements in the octahedral complex. Indeed, they may participate in π-bonding by overlapping sidewise with π-orbitals of the same six ligands that are approaching the central cation from both ends along each of the Cartesian axes x, y, and z. Despite their general weakness as compared to σ-bonds, the π-bonds nevertheless are significant in interpreting the distortions in symmetry from the basic structural shapes or arrangements for which the σ bonds are responsible. In their influence in shortening the internuclear distances of bonds, in modifying bond angles, and in promoting a tendency toward "planarity" upon an otherwise three-dimensional configuration, π-bonds become of greater importance in the square planar and tetrahedral complexes.

SEQUENCE OF ENERGY SEPARATIONS IN OCTAHEDRAL COMPLEX

We concentrate herein upon the octahedral complex. As already stated in the expositions of diatomic MO's, two molecular orbitals are always formed when two atomic orbitals overlap. One of these is antibonding, and of an energy level always higher than either of the two atomic orbitals from which it was formed. The other is bonding, and of an energy level always below that of the parent atomic orbitals. Consequently, the six valence orbitals of the central transition-metal ion in their overlapping with the six ligand σ valence orbitals (characteristic of octahedral coordination) must inevitably form six MO bonding orbitals and six MO antibonding orbitals. Adding to these twelve orbitals the three nonbonding d_{xy}, d_{xz}, and d_{yz} orbitals, we observe that the *aufbau* of the MO complexing process makes a total of fifteen orbitals potentially available for electron filling regardless of their specific classifications.

We may now suggest a few rules in accordance with which the sequence of their relative energy levels or separations may be denoted for the octahedral complex:

1. The six bonding orbitals of the molecule are not all equivalent in energy; they separate into three different groups of degenerated σ MO's. Although this separation can be reliably determined by measurements of symmetry, the particular *sequence* of these three bonding energy levels has yet to be established. Consequently, any arrangement of the three levels of bonding provided for purposes of illustration must be regarded as purely arbitrary. Actually, their exact order proves hardly more than academic here, inasmuch as all of these σ MO bonding orbitals must be fully occupied by electrons in any octahedral complex; hence, the properties of the complex with respect to bond order and bond type remain unaffected by the variability of their respective positions.

Suffice it, that the three MO bonding sets may be described adequately by the numbers and types of their respective orbital components. These are

(a) *singly degenerate* (one MO in the level). We designate this as σ_s to denote the involvement in its construction (in association with a σ ligand valence orbital) of the s valence orbital of the transition-metal ion.

(b) *doubly degenerate* (two MO's in the level). We designate this as σ_d to denote the involvement in their construction (in association with σ ligand valence orbitals) of the d_{z^2} and $d_{x^2-y^2}$ orbitals of the transition-metal ion.

(c) *triply degenerate* (three MO's in the level). We designate this as σ_p to denote the involvement in their construction (in association with σ ligand valence orbitals) of the p_x, p_y, and p_z orbitals of the transition-metal ion.

2. The six antibonding orbitals are, likewise, only partly degenerate. Their relative degrees of orbital degeneracy may also be described as *singly* degenerate (σ_s^*), *doubly* degenerate (σ_d^*), and *triply* degenerate (σ_p^*) antibonding energy sequences. As the relative levels of these three sets of antibonding orbitals have been reliably established for the ground states of complexes of conventional familiarity, their assigned order may well be different from that arbitrarily given to their analogs of the bonding set.

3. The three nonbonding orbitals of the central transition-metal ion — d_{xy}, d_{xz}, d_{yz} — occupy an energy level intermediate between the bonding and antibonding orbital groups. They are thus more stable than the antibonding d_{z^2} and $d_{x^2-y^2}$ orbitals (σ_d^*) but less stable than the bonding d_{z^2} and $d_{x^2-y^2}$ orbitals (σ_d). We refer to this nonbonding set as π_d to denote their potential availability to the overlap construction of π MO's with ligands capable of forming π-bonds. The accommodating of electrons in these orbitals is restricted to electrons of the transition-metal ion.

4. Orderly build-up of MO configuration begins with simply computing the total number of electrons involved in the formation of the complex. This obviously comprises the twelve always supplied by the sigma orbitals of the six ligands in octahedral complexing (regardless of their specific chemical identities, two from each) plus the specific number of electrons in the d orbitals of the transition-metal ion, of variable identity.

Irrespective of source — transitional or ligand species — twelve electrons must first be accommodated in the six MO σ bonding levels to fulfil octahedral orientation. We then reach the level of the three degenerate nonbonding π MO orbitals. These must be carefully appraised in relation to their possibly near equivalences of energy with the very lowest antibonding level — that of the two antibonding d_{z^2} and $d_{x^2-y^2}$ orbitals. With weakly covalent bondings between the ligands and the central transition-metal ion that involve relatively small overlap, the energy of the level is so slightly removed from that of the π_d level that the joint occupation of their respective orbitals by unpaired electrons is rendered feasible even before electrons pair in the lower level. In weak overlap, therefore, each orbital of both the lower triply degenerate π_d level and the higher doubly degenerate σ_d^* level receives one electron apiece, with parallel spins, before pairing occurs in the lower π_d level. In strongly covalent bondings that involve relatively large overlap, the upper σ_d^* level is sufficiently removed from the π_d level to render the former highly unstable with respect to the latter. Consequently, the electrons in the σ_d^* level cannot be energetically accommodated until *after* all orbitals of the lower π_d level have received their individual maximum quotas of two paired electrons.

One should require no prodding to be reminded of the immediate resemblances of these systematizations to the weak fields (high spins) and strong fields (low spins) of the electrostatic crystal field theory. Except for the differences in approach, they are qualitatively identical; the triply degenerate MO π_d level of d_{xy}, d_{xz}, and d_{yz} orbitals is merely the t_{2g} level of the electrostatic field; and the doubly degenerate MO σ_d^* level of d_{z^2} and $d_{x^2-y^2}$ orbitals the e_g level of the electrostatic field.

Figure 4.4 is a generalized diagram of molecular orbitals which illustrates the preceding rules for the octahedral system, whence the illustrations to follow draw their interpretations. It is to be stressed that the degrees of

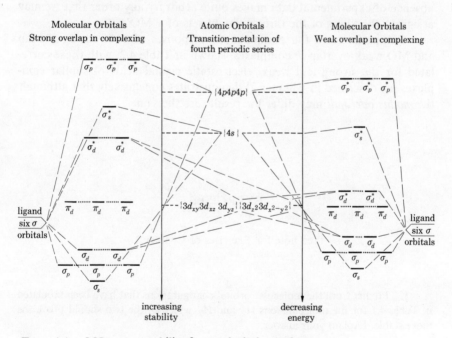

FIG. 4.4   MO energy-stability for octahedral complexes.

energy separations of the orbital levels are not intended to suggest actual quantitative measurement; they are purely qualitatively descriptive with respect to our present needs.

The derivations of MO configurations which we now make to demonstrate the applications of these energy-level diagrams are greatly simplified by the ordered sequence of MO electron-filling that they express, as follows:

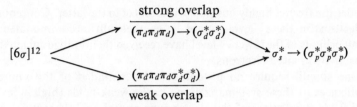

In this diagram, the designation $[6\sigma]^{12}$ represents the six sigma-bonding MO's — regardless of how they are constructed — all completely filled with their individual maximum quotas of two electrons (hence, the superscript 12).

Such collation is purely academic, because, in octahedral complexing, these bonding σ MO's are never involved in the evaluating of paramagnetism and other properties attributable to the unpaired electrons frequently to be found in MO's of higher energy. Moreover, as we have already noted, the absence of experimental data makes quite arbitrary any order that we may assign to the filling of the three different sets of σ MO's.

Comparisons of the paramagnetic responses of MO strong-overlap and MO weak-overlap d^ϵ complexes, shown in Table 4.2, with those correlated for the strong and weak electrostatic crystal fields of similar complexes (as described in Chapter 2) demonstrates conclusively that although the *modus operandi* may differ the results are the same.

EXERCISES

(See note for Exercises of Chapter 1.)

1. Predict from the molecular orbital configurations that have been tabulated in Table 4.1 for the molecule-ions H_2^+ and H_2^- which of the two should prove the more stable. Explain your answer.

2. The most recent experimental evidence of the lack of paramagnetic response on the part of the C_2 molecule supports the description of its ground state as

$$KK\sigma_{2s}^2\sigma_{2s}^{*2}(\pi_{2p}^2\pi_{2p}^2);$$

this means specifically that the two degenerate π_{2p} orbitals are filled *before* the σ_{2p} rather than *after*, as previously held. Nonetheless, the past uncertainties of the

Table 4.2

Central metal Transition d^e ion (e = total d electrons)	MO electrons total = $(12 + e)$	Strong Overlap MO Configuration	Unpaired electrons	Weak Overlap MO Configuration	Unpaired electrons
d^1	13	$[6\sigma]^{12}\pi_d^1$	1	$[6\sigma]^{12}\pi_d^1$	1
d^2	14	$[6\sigma]^{12}\pi_d^1\pi_d^1$	2	$[6\sigma]^{12}\pi_d^1\pi_d^1$	2
d^3	15	$[6\sigma]^{12}\pi_d^1\pi_d^1\pi_d^1$	3	$[6\sigma]^{12}\pi_d^1\pi_d^1\pi_d^1$	3
d^4	16	$[6\sigma]^{12}\pi_d^2\pi_d^1\pi_d^1$	2	$[6\sigma]^{12}\pi_d^1\pi_d^1\pi_d^1\sigma_d^{*1}$	4
d^5	17	$[6\sigma]^{12}\pi_d^2\pi_d^2\pi_d^1$	1	$[6\sigma]^{12}\pi_d^1\pi_d^1\pi_d^1\sigma_d^{*1}\sigma_d^{*1}$	5
d^6	18	$[6\sigma]^{12}\pi_d^2\pi_d^2\pi_d^2$	0	$[6\sigma]^{12}\pi_d^2\pi_d^1\pi_d^1\sigma_d^{*1}\sigma_d^{*1}$	4
d^7	19	$[6\sigma]^{12}\pi_d^2\pi_d^2\pi_d^2\sigma_d^{*1}$	1	$[6\sigma]^{12}\pi_d^2\pi_d^2\pi_d^1\sigma_d^{*1}\sigma_d^{*1}$	3
d^8	20	$[6\sigma]^{12}\pi_d^2\pi_d^2\pi_d^2\sigma_d^{*1}\sigma_d^{*1}$	2	$[6\sigma]^{12}\pi_d^2\pi_d^2\pi_d^2\sigma_d^{*1}\sigma_d^{*1}$	2
d^9	21	$[6\sigma]^{12}\pi_d^2\pi_d^2\pi_d^2\sigma_d^{*2}\sigma_d^{*1}$	1	$[6\sigma]^{12}\pi_d^2\pi_d^2\pi_d^2\sigma_d^{*2}\sigma_d^{*1}$	1

relative positions of these orbital levels invite our speculations as to the nature of the differences the molecule would manifest, in contrast with the properties already tabulated for it in this chapter, were these orbital levels to be transposed.

If such an interchange of orbitals were then made, describe the orbital, the bond order, bond type, and degrees of paramagnetic response of the molecule. Assume that all the rules of mathematical regularity pertinent to electron-filling are scrupulously observed.

3. Write the molecular orbital configurations of the respective molecule-ions that are formed by removing just one electron from each of the following neutral molecules:
(a) the oxygen molecule, O_2.
(b) the nitrogen molecule, N_2.

4. Determine for each of the molecule-ions obtained in the preceding exercise its net bond order and number of unpaired electron spins. Assume that significant σ–σ interaction occurs.

5. The N_2^+ and O_2^+ molecule-ions have the same net bond order and identical numbers of electrons in bonding orbitals. The bond-dissociation energy of N_2^+, however, is a bit larger than that of O_2^+, indicating that the former is somewhat the more stable of the two. Account for this.

6. The homonuclear diatomic molecules of both the alkali-metal and halogen families of elements are all diamagnetic. Predict, in plausible consistency with mathematical regularity of ground state assignments of electrons to orbitals, the molecular configurations, bond orders, and bond types of
(a) Na_2, K_2, Rb_2, Cs_2
(b) F_2, Cl_2, Br_2, I_2
Assume that the energy separations of these halogen molecules cause the sigma-bonding orbitals of the valence levels to be filled before those of their pi levels.

7. Because of the differences in the degree and order of the electron-filling of orbitals of identical main energy levels of atoms, it does not follow that a molecular shell of one species necessarily has the same electron population as a similar shell of another. For example, the NN shell of Rb_2 contains precisely 16 electrons (inasmuch

as the N shell of each of the two contributory Rb^0 atoms has the electronic configuration $4s^2\ 4p^6$), whereas the NN shell of Cs_2 contains a total of 36 electrons, each Cs^0 atom conforming to an electronic configuration of $4s^2\ 4p^6\ 4d^{10}$.

In consistency with the ground state filling of atomic orbitals, supply the total electron content of the following molecular shells:
   (a) the MM of Rb_2
   (b) the OO of Cs_2
   (c) the MM of I_2
   (d) the NN of I_2
   (e) the LL of Rb_2

8. The existence of the NO^+ molecule-ion is well established. Provide an electron-dot configuration for this heterodiatomic species that is consistent with its MO description, net bond order, and bond type.

9. A bond order of 3 is computed for the molecule AB. Is it possible that, in the absence of further information, this triple bond can check out as two σ- and one π? Explain your answer.

10. With respect to their energy and their spatial orientation, what resemblances and differentiations exist between the individually separate orbital components of either the π levels or the π^* levels of molecules?

11. A *free radical* is defined as a molecule or molecule-ion having one or more electrons of unpaired spin; hence, it is a paramagnetic species of permanent dipole moment. Support with formulated MO descriptions your conclusions as to which, if either, of the molecules, CO and O_2, is
   (a) a free radical.
   (b) a *di*radical (a free radical possessed of two unpaired electrons).
   (c) not a free radical.

APPENDIX A: Answers to Exercises

CHAPTER 1. Atomic Properties

1. -2.177×10^{-11} erg/atom; -313.5 kcal/mole

2. $E = E_2 - E_1 = \dfrac{2\pi^2 m Z^2 e^4}{h^2}\left(\dfrac{1}{n_1^2} - \dfrac{1}{n_2^2}\right)$

3. (a) $R = 2\pi^2\, me^4/h^3 c$
   (b) (1) $R = 1.097 \times 10^5$ cm^{-1}; (2) $R = 1.097 \times 10^{-3}$ Å^{-1}

4. (a) $\lambda = 911.6$Å
   (b) 3646.4Å

5. (a) 2.179×10^{-11} erg/atom
   (b) 313.5 kcal/mole
   (c) 13.60 eV/atom

6. (a) $n = 2$; -5.44×10^{-12} erg/atom $= -78.3$ kcal/mole
       $n = 3$; -2.42×10^{-12} erg/atom $= -34.8$ kcal/mole
       $n = 4$; -1.36×10^{-12} erg/atom $= -19.6$ kcal/mole
       $n = 5$; -0.871×10^{-12} erg/atom $= -12.5$ kcal/mole
   (b) a maximum value of zero (see data, Exercise 1)

7. Lyman: $[n_2 = 2] \to [n_1 = 1]$, λ first line $= 1215$Å
   Balmer: $[n_2 = 3] \to [n_1 = 2]$, λ first line $= 6563$Å
   Ritz-Paschen: $[n_2 = 4] \to [n_1 = 3]$, λ first line $= 18{,}750$Å
   Brackett: $[n_2 = 5] \to [n_1 = 4]$, λ first line $= 40{,}510$Å
   Pfund: $[n_2 = 6] \to [n_1 = 5]$, λ first line $= 74{,}580$Å

8. With $n_2 < n_1$, the wave number $\bar{\nu}$, denoting the number of waves/cm, would be a completely meaningless negative quantity; and with $n_2 = n_1$, there could be no spectral line inasmuch as $\bar{\nu}$ would be zero.

9. (a) In each instance there is just a single orbital electron.
   (b) Frequencies and wave numbers vary with the square of the atomic number. Consequently, those of He$^+$ are $(2)^2 = 4$ times greater; those of Li^{2+} are

149

150   APPENDIX A

$(3)^2 = 9$ times greater; those of Be^{3+} are $(4)^2 = 16$ times greater than those of H^0.

(c) Wavelengths vary inversely with the square of the atomic number. Consequently, those of He^+, Li^{2+}, and Be^{3+} are one-fourth, one-ninth, and one-sixteenth, respectively, of the corresponding dimensions of the H^0.

(d) $[n_2 = 3] \to [n_1 = 2]$ $\begin{cases} \lambda \text{ first line } He^+ = 1641\text{Å} \\ \lambda \text{ first line } Li^{2+} = 729\text{Å} \\ \lambda \text{ first line } Be^{3+} = 410\text{Å} \end{cases}$

10. Orbitals are most stable when all in a given subshell are either completely filled (two electrons in each), or just half-filled (one electron in each).

11. Because of their mutual repulsions of charge, electrons are more stable in separate orbitals than together in the same orbital. As each electron, in effect, is a tiny spinning magnet, it becomes paired, in the absence of separate orbital accommodations of requisite energies, by mutual associations of electrons of opposing spin.

12. (a) $Cu^0 = 3d^2\ 3d^2\ 3d^2\ 3d^2\ 3d^2\ 4s^1$: spherical
   $Cu^+ = 3d^2\ 3d^2\ 3d^2\ 3d^2\ 3d^2\ 4s^0$: spherical
   $Cu^{2+} = 3d^2\ 3d^2\ 3d^2\ 3d^2\ 3d^1\ 4s^0$: nonspherical
  (b) $N^0 = 2p^1\ 2p^1\ 2p^1$: spherical
   $N^{3-} = 2p^2\ 2p^2\ 2p^2$: spherical
  (c) $Mn^0 = 3d^1\ 3d^1\ 3d^1\ 3d^1\ 3d^1\ 4s^1$: spherical
   $Mn^{2+} = 3d^1\ 3d^1\ 3d^1\ 3d^1\ 3d^1\ 4s^0$: spherical
   $Mn^{3+} = 3d^1\ 3d^1\ 3d^1\ 3d^1\ 3d^0\ 4s^0$: nonspherical
  (d) $Fe^{2+} = 3d^2\ 3d^1\ 3d^1\ 3d^1\ 3d^1\ 4s^0$: nonspherical
   $Fe^{3+} = 3d^1\ 3d^1\ 3d^1\ 3d^1\ 3d^1\ 4s^0$: spherical
  (e) $Ga^0 = 4s^2\ 4p^1$: nonspherical
   $Ga^{3+} = 4s^0\ 4p^0$: spherical
  (f) $Ne^0 = 2s^2\ 2p^2\ 2p^2\ 2p^2$: spherical
   $Ne^+ = 2s^2\ 2p^2\ 2p^2\ 2p^1$: nonspherical
   $Ne^{3+} = 2s^2\ 2p^1\ 2p^1\ 2p^1$: spherical

13. $He^+ = 54.40$ eV; $Li^{2+} = 122.4$ eV; $Be^{3+} = 217.6$ eV; $B^{4+} = 340.0$ eV; $C^{5+} = 489.6$ eV

14. (a) $He^+_{(g)} \to He^{2+}_{(g)} + 1e^-$: (2nd ionization potential of He^0)
   $Li^{2+}_{(g)} \to Li^{3+}_{(g)} + 1e^-$: (3rd ionization potential of Li^0)
   $Be^{3+}_{(g)} \to Be^{4+}_{(g)} + 1e^-$: (4th ionization potential of Be^0)
   $B^{4+}_{(g)} \to B^{5+}_{(g)} + 1e^-$: (5th ionization potential of B^0)
   $C^{5+}_{(g)} \to C^{6+}_{(g)} + 1e^-$: (6th ionization potential of C^0)
  (b) -7840 kcal/mole

15. The addition of a single electron to the neutral nitrogen atom (N^0) must, perforce, contend with the stability it already has by virtue of its precisely half-filled orbital subshell of $2p^1\ 2p^1\ 2p^1$. Hence, the single electron is not accepted. There is no alternate to the stability the atom already has other

APPENDIX A   151

than the acceptance of three electrons, to give the N^{3-} ion ($2p^2\ 2p^2\ 2p^2$). By definition, electron affinity concerns the addition of a *single* electron to the neutral atom. As the configurations of the neutral carbon and oxygen atoms, on the other hand, are still short of stability their respective subshells may accept one electron each, as follows:

$$[2p^1\ 2p^1\ 2p^0\ (= C^0)] + 1e^- \rightarrow [2p^1\ 2p^1\ 2p^1(= C^-)]$$

and

$$[2p^2\ 2p^1\ 2p^1\ (= O^0)] + 1e^- \rightarrow [2p^2\ 2p^2\ 2p^1\ (= O^-)].$$

16. (a) The volume of a sphere is defined by $v = \frac{4}{3}\pi r^3$; hence,

$$\text{nuclear volume} = \frac{4}{3} \times 3.1416 \times (1.6 \times 10^{-13})^3$$
$$= 1.72 \times 10^{-38} \text{cc}$$

(b) atomic diameter $= 2r = 2\sqrt[3]{\frac{3v}{4\pi}}$, ($v$ = atomic volume)

$$= 2\sqrt[3]{\frac{3 \times 1.14 \times 10^{-22}}{4 \times 3.1416}}$$
$$= 6.02 \times 10^{-8} \text{cm}$$

(c) $\text{ratio} = \frac{1.60 \times 10^{-13} \text{cm}}{3.01 \times 10^{-8} \text{cm}} = \frac{5.32 \times 10^{-6}}{1}$

(d) $\frac{1.72 \times 10^{-38} \text{cc}}{1.14 \times 10^{-22} \text{cc}} \times 100 = 1.51 \times 10^{-16}\%$ of space.

17. (a) $\lambda = \frac{6.63 \times 10^{-27} \text{ erg} \cdot \text{sec}}{(9.11 \times 10^{-28} \text{gm}) \times (4.60 \times 10^8 \text{cm/sec})}$
$$= 1.58 \times 10^{-8} \text{ erg} \frac{\text{sec}^2}{\text{gm cm}}$$

and since 1 erg = 1 gm cm^2/sec^2

$$\lambda = 1.58 \times 10^{-8} \frac{\text{gm cm}^2}{\text{sec}^2} \times \frac{\text{sec}^2}{\text{gm cm}} = 1.58 \times 10^{-8} \text{ cm}$$

(b) $\lambda = \frac{h}{e_m c}$

$$= \frac{6.63 \times 10^{-27} \text{ erg}\cdot\text{sec}}{(9.11 \times 10^{-28} \text{ gm}) \times (3.00 \times 10^{10} \text{ cm/sec})}$$
$$= 2.43 \times 10^{-10} \text{ erg} \frac{\text{sec}^2}{\text{gm cm}} = 2.43 \times 10^{-10} \text{ cm}$$

18. (a) The Einstein equation relates mass to energy as $E = mc^2$. The substitutions and conversions to be made therein respond, consequently, as follows:

$$E_{He}^0 = \left(4.003 \text{ amu} \times \frac{1.66 \times 10^{-24} \text{ gm}}{1 \text{ amu}}\right) \times \left(3.00 \times 10^{10} \frac{\text{cm}}{\text{sec}}\right)^2$$
$$= 5.98 \times 10^{-3} \text{ gm cm}^2/\text{sec}^2.$$

But, gm cm^2/sec^2 are the dimensions of the erg; therefore

$$5.98 \times 10^{-3} \text{ erg} \times \frac{2.39 \times 10^{-8} \text{ cal}}{1 \text{ erg}} = 1.43 \times 10^{-10} \text{ cal.}$$

152   APPENDIX A

(b) 5.98×10^{-3} erg $\times \dfrac{1 \text{ eV}}{1.60 \times 10^{-12} \text{ erg}} \times \dfrac{1 \text{ MeV}}{10^6 \text{ eV}}$

yields 3.74×10^3 MeV.

19. The number of energy sublevels, sequentially, $s, p, d, f, g, h \ldots$, cannot be greater than that numerically defined by the principal quantum number. Therefore, the orbitals $2d$, $3f$, $4g$, and $5h$ must be excluded.

20. (a) 72 electrons
    (b) 18 electrons

CHAPTER 2.   Properties of Chemical Bond

1. (a) To form the third S–F bond required for SF_3, another half-filled (one-electron) orbital must be created. But, any promotion of an electron from the $3p$ electron-pair ($S^0 = 3s^2\ 3p^2\ 3p^1\ 3p^1$) to a $3d$ orbital inevitably creates two additional valence orbitals ((i.e., $3s^2\ 3p^1\ 3p^1\ 3p^1\ 3d^1$). Consequently, SF_4 forms instead.
Similarly, the formation of SF_5 would require the availability of five valence orbitals around the sulfur atom. Any enlargement beyond four would obviously require promotion of an electron from the $3s$ electron-pair of the sulfur atom, as well as from a $3p$ electron-pair; that is, the configuration, $3s^1\ 3p^1\ 3p^1\ 3p^1\ 3d^1\ 3d^1$ would necessarily result. Consequently, SF_6 forms instead.

(b) The orbital distribution of the valence electrons of the nitrogen atom is $2s^2\ 2p^1\ 2p^1\ 2p^1$. The three half-filled orbitals that it supplies permit the normal expected formation of three covalent bonds with three chlorine atoms ($\times Cl\overset{\times\times}{\underset{\times\times}{\times}}$). In order to form NCl_5 one of the paired electrons of the $2s$ orbital would have to be promoted to the very next principal quantum level ($n = 3$). Such a transition is, apparently, prohibited by stability requirements.

(c) The valence-shell of xenon may be represented by $\left[\overset{\times\times}{\underset{\times\times}{\times Xe \times}} \right]$, corresponding to [Kr] $4d^{10}\ 5s^2\ 5p^2\ 5p^2\ 5p^2$. We have available the five completely vacant $6d$ orbitals. Promotion of an electron from one of the $5p$ electron-pairs opens up two bonding accommodations (single-electron orbitals). Promotion of an additional electron from a second $5p$ electron-pair makes available four bonding accommodations (four single-electron orbitals). The similar promotion of a third electron from a third $5p$ electron-pair makes possible a total of six Xe-F bonds. This would conform, consequently, to the orbital descriptions of the xenon atom in the various compounds, as follows:

APPENDIX A   153

$XeF_2 \left(= \overset{xx}{\underset{xx}{\times}} Xe \overset{F}{\underset{F}{<}} \right)$, $5s^2\ 5p^2\ 5p^2\ 5p^1\ 6d^1$ ———→available to 2 $\left[\overset{xx}{\underset{xx}{\times}} F \times \right]$

10 electrons on Xe

$XeF_4 \left(= \overset{F}{\underset{F}{>}} Xe \overset{xx}{\underset{xx}{<}} \overset{F}{\underset{F}{}} \right)$, $5s^2\ 5p^2\ 5p^1\ 5p^1\ 6d^1\ 6d^1$ ———→available to 4 $\left[\overset{xx}{\underset{xx}{\times}} F \times \right]$

12 electrons on Xe

$XeF_6 \left(= \begin{array}{c} F \\ \diagdown \\ F \end{array} \overset{F \underset{xx}{\diagup} F}{\underset{|}{\overset{}{Xe-F}}} \right)$, $5s^2\ 5p^1\ 5p^1\ 5p^1\ 6d^1\ 6d^1\ 6d^1$ ———→available to 6 $\left[\overset{xx}{\underset{xx}{\times}} F \times \right]$

14 electrons on Xe

2. $CHF_3 > CHCl_3 > CHBr_3 > CH_2Br_2 > CH_4$

The most electronegative element shown in any of the four carbon linkages common to all of the compounds is fluorine; next is chlorine; thereafter bromine; and least electronegative is hydrogen. The greater attractions of the fluorine atoms for the electron-pairs they share with the single carbon atom causes, as a concomitant or parallel effect, a greater attraction by the carbon atom for the electron-pairs it shares with the hydrogen. In essence, this means a general displacement of the electron density away from the hydrogen atom. The greater this displacement, the greater becomes the increment of positive charge on the hydrogen atom and, also, the greater the increment of negative charge upon the carbon atom. Consequently, the C–H bond becomes more polar.

3. (a) 3.01×10^{23} quanta; 0.500 einstein

(b) Energy absorption for the breaking of an Avogadro number (N) of bonds conforms to the formulation $E = Nh\nu$. Consolidating this with $\lambda\nu = c$ yields

$$\lambda = N \left[\frac{hc}{E} \right].$$

Substitution of required constants and conversion units in this latter expression leads to

$\lambda = 6.02 \times 10^{23}$ bonds $\times$

$$\left[\frac{(6.63 \times 10^{-27}\ \text{erg/sec}) \times (3.00 \times 10^{10}\ \text{cm/sec})}{(8.03 \times 10^4\ \text{cal/bond}) \times (4.18\ \text{joule/cal} \times 10^7\ \text{ergs/joule})} \right]$$

$= 3.57 \times 10^{-5}$ cm.

(c) The frequency, ν, and the wave number, $\bar{\nu}$, of the required light radiation conform to

$$\nu = \frac{3.00 \times 10^{10}\ \text{cm/sec}}{3.57 \times 10^{-5}\ \text{cm}} = 8.40 \times 10^{14}\ \text{sec}^{-1}$$

$$\bar{\nu} = \frac{1}{\lambda} = \frac{1}{3.57 \times 10^{-5}\ \text{cm}} = 2.80 \times 10^4\ \text{waves/cm}$$

4. (a) $NaF > LiBr > NaCl > RbF > KI > CsI$

(b) (1) Largest, LiF; (2) smallest, CsI

5. (a) fc H = 0; fc N = 0
   (b) fc B = −1; fc H = 0; fc F = 0; fc N = +1
   (c) fc N = −1; fc C = 0; fc O = 0
   (d) fc H = 0; fc C = 0; fc N = 0
   (e) fc Cl = 0; fc S = 0
   (f) fc H = 0; fc N_α = 0; fc N_β = +1; fcN_γ = −1
   (g) fc H = 0; fc C = 0; fc O = 0
   (h) fc B = −1; fc F = 0

6.

\[structure drawings of ozone resonance forms\]

7. (a) , 22 electrons
   (b) 6 unshared electrons; 3 unshared electron-pairs; 2 electron-pair bonds
   (c) zero

CHAPTER 3. Complexing Transitional Metals

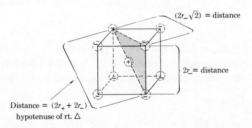

FIG. A.1   Reference cube (body-centered) for calculation of a radius ratio (*R*).

1.
$$(2r_+ + 2r_-)^2 = (2r_-\sqrt{2})^2 + (2r_-)^2$$
$$4r_+^2 + 8r_+r_- + 4r_-^2 = 8r_-^2 + 4r_-^2$$
$$4r_+^2 + 8r_+r_- - 8r_-^2 = 0$$

$$\left(\frac{r_+}{r_-}\right)^2 + 2\left(\frac{r_+}{r_-}\right) - 2 = 0; \text{ whence, by the quadratic we obtain}$$

$$R = \frac{r_+}{r_-} = \frac{-2 \pm \sqrt{12}}{2} = \frac{-2 \pm 3.464}{2} = \frac{1.464}{2} = 0.732 \text{ acceptable root}$$

2. (a) *R* = 0.85, which is very much greater than the value of 0.732 representing the lower limit of 8-coordination; hence, the structure is cubic.

(b) $R = 0.21$, which is less than the value of 0.225 representing the lower limit of tetrahedral structure. The inexactitudes of radius measurements do not permit sufficiently precise demarcations for predicting borderline values of R. These values can represent only guide lines to predictions.

(c) $R = 0.33$, which is very much less than the value of 0.414 representing the lower limit of octahedral structure. Here, the fairly large divergence from guide-line predictions of R would suggest a significant degree of covalent bonding between the anions. Covalence means overlap and, consequently, increased opportunities for squeezing in additional ligands. Six would be accommodated here instead of the expected four.

This situation of intimate ligand-ligand contacts arising from enlarged coordination number requires some logical reconciliation with the now-increased electron-cloud repulsions of the overcrowded ligands. Apparently, the balancing compensations required for stability are made by withdrawal of the ligands away from contact with the central atom. Illustratively, the following top-view cross-section of the octahedron of LiCl would suggest the manner in which such adjustments are made:

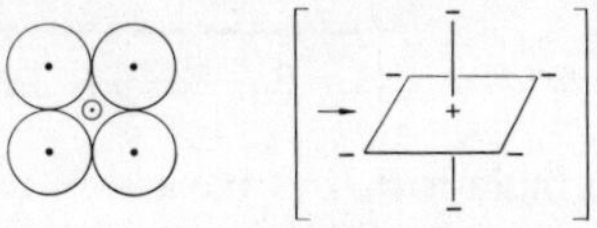

FIG. A.2   Reference cross-section of LiCl octahedron.

We may plausibly conclude therefrom, then, that the octahedral-6 arrangement of the LiCl crystal is being maintained by ligand-ligand contact rather than by cation-ligand contact.

3. In the CoF_6^{3-} ion the ligand-splitting of the orbitals of the central meltallic ion (Co(III) is a d^6 ion) is small enough to make the higher e_g level favorable to electron occupation despite the availability of only partly filled orbitals of the t_{2g} lower level. Hence, CoF_6^{3-} is in a high-spin state of $(t_{2g})^4(e_g)^2$. This corresponds to four unpaired electrons. In the ammine complex, however, the ligand-splitting of the Co(III) orbitals is sufficiently large to render occupation of the higher-energy e_g level energetically unfavorable as long as accommodations exist in orbitals of the lower level. Consequently, there is increased pairing of electrons in the lower t_{2g} set resulting in the low-spin state of $(t_{2g})^6(e_g)^0$. There then is zero unpaired electrons, and the species proves diamagnetic.

4. (a) Cr(III) is a d^3 central species; hence, orbital compositions are $(t_{2g})^3(e_g)^0$
   V(III) is a d^2 central species; hence, orbital compositions are $(t_{2g})^2(e_g)^0$
   Ti(III) is a d^1 central species; hence, orbital compositions are $(t_{2g})^1(e_g)^0$

   (b) Cr(I) = $(t_{2g})^3(e_g)^2$ high spin, 5 unpaired electrons
       = $(t_{2g})^5(e_g)^0$ low-spin, 1 unpaired electron
       Cr(II) = $(t_{2g})^3(e_g)^1$ high-spin, 4 unpaired electrons
       = $(t_{2g})^4(e_g)^0$ low-spin, 2 unpaired electrons

5. (a) Non-carbonyl Ni^0 = $3d^2\ 3d^2\ 3d^2\ 3d^1\ 3d^1\ 4s^2\ 4p^0\ 4p^0\ 4p^0$

Carbonyl Ni^0 = $3d^2\ 3d^2\ 3d^2\ 3d^2\ 3d^2$ $\begin{bmatrix} CO & CO & CO & CO \\ xx & xx & xx & xx \\ 4s & 4p & 4p & 4p \end{bmatrix}$

Hence, the formula is $Ni(CO)_4$; orbital hybridization is sp^3; arrangement is tetrahedral.

(b) Non-carbonyl Fe^0 = $3d^2\ 3d^1\ 3d^1\ 3d^1\ 3d^1\ 4s^2\ 4p^0\ 4p^0\ 4p^0$

Carbonyl Fe^0 = $3d^2\ 3d^2\ 3d^2\ 3d^2$ $\begin{bmatrix} CO & CO & CO & CO & CO \\ xx & xx & xx & xx & xx \\ 3d & 4s & 4p & 4p & 4p \end{bmatrix}$

Hence, the formula is $Fe(CO)_5$; orbital hybridization is dsp^3; arrangement is trigonal bipyramidal.

(c) Non-carbonyl Cr^0 = $3d^1\ 3d^1\ 3d^1\ 3d^1\ 3d^1\ 4s^1\ 4p^0\ 4p^0\ 4p^0$

Carbonyl Cr^0 = $3d^2\ 3d^2\ 3d^2$ $\begin{bmatrix} CO & CO & CO & CO & CO & CO \\ xx & xx & xx & xx & xx & xx \\ 3d & 3d & 4s & 4p & 4p & 4p \end{bmatrix}$

Hence, the formula is $Cr(CO)_6$; orbital hybridization is d^2sp^3; arrangement is octahedral.

6. (a) Non-carbonyl Co^0 = $3d^2\ 3d^2\ 3d^1\ 3d^1\ 3d^1\ 4s^2\ 4p^0\ 4p^0\ 4p^0$

$\left.\begin{array}{l}\text{Carbonyl } Co^0 \\ \text{Carbonyl } Co^0\end{array}\right\} \begin{array}{l}Co\text{—}Co \\ bridge\end{array} \left\{\begin{array}{l} = 3d^2\ 3d^2\ 3d^2\ 3d^2\ 3d^1 \\ \longleftrightarrow \\ = 3d^2\ 3d^2\ 3d^2\ 3d^2\ 3d^1\end{array}\right.$ $\begin{bmatrix} CO & CO & CO & CO \\ xx & xx & xx & xx \\ 4s & 4p & 4p & 4p \end{bmatrix}$ $\downarrow\uparrow$ $\begin{bmatrix} CO & CO & CO & CO \\ xx & xx & xx & xx \\ 4s & 4p & 4p & 4p \end{bmatrix}$

Hence, the formula is that of the *dimer*, with bridging by pairing of the two $3d^1$ electrons of opposing spins, to yield $[Co(CO)_4]_2 = Co_2(CO)_8$

Non-carbonyl Re^0 = $4f^{14}\ 5d^1\ 5d^1\ 5d^1\ 5d^1\ 5d^1\ 6s^2\ 6p^0\ 6p^0\ 6p^0$

$\left.\begin{array}{l}\text{Carbonyl } Re^0 \\ \text{Carbonyl } Re^0\end{array}\right\} \begin{array}{l}Re\text{-}Re \\ bridge\end{array} \left\{\begin{array}{l} = 4f^{14}\ 5d^2\ 5d^2\ 5d^2\ 5d^1 \\ \longleftrightarrow \\ = 4f^{14}\ 5d^2\ 5d^2\ 5d^2\ 5d^1\end{array}\right.$ $\begin{bmatrix} CO & CO & CO & CO & CO \\ xx & xx & xx & xx & xx \\ 5d & 6s & 6p & 6p & 6p \end{bmatrix}$ $\downarrow\uparrow$ $\begin{bmatrix} CO & CO & CO & CO & CO \\ xx & xx & xx & xx & xx \\ 5d & 6s & 6p & 6p & 6p \end{bmatrix}$

Hence, the formula is that of the *dimer*, with bridging by pairing of the two $5d^1$ electrons of opposing spins, to yield $[Re(CO)_5]_2 = Re_2(CO)_{10}$

(b) Like rhenium, the element manganese has seven electrons that respond to rearrangement, the orbital configuration of the Mn^0 being $3d^5\ 4s^2$. Consequently, the carbonyl of the metal must conform to the formulation of the dimer $[Mn(CO)_5]_2 = Mn_2(CO)_{10}$

7. As the coordination number of the vanadium in this neutral complex is six, the arrangement of the complex is octahedral; that is, the orbital hybridization is d^2sp^3. Hence,

Non-carbonyl V^0 = $3d^1\ 3d^1\ 3d^1\ 3d^0\ 3d^0\ 4s^2\ 4p^0\ 4p^0\ 4p^0$

Carbonyl V^0 = $3d^2\ 3d^2\ 3d^1$ $\begin{bmatrix} CO & CO & CO & CO & CO & CO \\ xx & xx & xx & xx & xx & xx \\ 3d & 3d & 4s & 4p & 4p & 4p \end{bmatrix}$

As there is one electron unpaired, the complex is thus paramagnetic rather than diamagnetic.

8. (a) $E = Nh\nu = \dfrac{2.859}{\lambda cm}$ cal/mole

$\lambda cm = \dfrac{2.859 \text{ cal/mole}}{42.9 \times 10^{-3} \text{ cal/mole}}$

$= 6.66 \times 10^{-5}$ cm

APPENDIX A 157

Hence, the color observed is the complementary color in this range; namely, green.

(b) $\lambda = \dfrac{c}{\nu} = \dfrac{3.00 \times 10^{10} \text{ cm/sec}}{5.3 \times 10^{14} \text{ sec}^{-1}}$
$= 5.64 \times 10^{-5}$ cm

Hence, the color of the absorbed visible light radiation is yellow.

9. The answer must be NO. The information does not permit an incontrovertible differentiation between stability (thermodynamic property measured by an equilibrium constant) and lability (kinetic property measured by rate of reaction). A substance may be stable yet not necessarily inert (non-labile). Consequently, the complex under increased temperature may be manifesting a greater lability, rather than a smaller stability, than the same complex under increased acidity.

10. $Ag_2[HgI_4] \leftarrow HgAg[AgI_4]$
    yellow   *red*

11. With dotted lines indicating the bidentate ligand attachments:

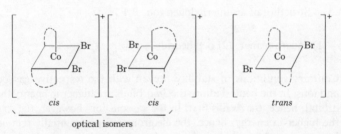

FIG. A.3 Optical and *cis-trans* isomers. Stereoisomerism of $[Co(en)_2Br_2]^+$.

12. $[Co(NH_3)_4ClI]Cl$

13. $\quad\quad\quad\quad [Cu(NH_3)_3Cl]^+[Pt(NH_3)Cl_3]^-$
    $\quad\quad\quad\quad [Pt(NH_3)_3Cl]^+[Cu(NH_3)Cl_3]^-$
    $\quad\quad\quad\quad [Pt(NH_3)_4]^{2+}[CuCl_4]^{2-}$

14. $[ONO-Co(NH_3)_5]^{2+}$ and $[(NH_3)_5Co-NO_2]^{2+}$
        nitrito isomer   nitro isomer

15. (a) $n = 2$
    $[Pd(NH_3)_3Cl]^+[Pd(NH_3)Cl_3]^-$
    $[Pd(NH_3)_4]^{2+}[PdCl_4]^{2-}$
    (b) $n = 3$
    $([Pd(NH_3)_3Cl]^+)_2[PdCl_4]^{2-}$
    $[Pd(NH_3)_4]^{2+}([Pd(NH_3)Cl_3]^-)_2$

16.

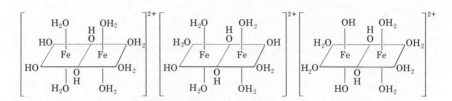

FIG. A.4 Isomeric forms of an hydroxy-bridged complex, $[Fe_2(OH)_4(H_2O)_6]^{2+}$.

17. (a) six electron-pairs
    (b)

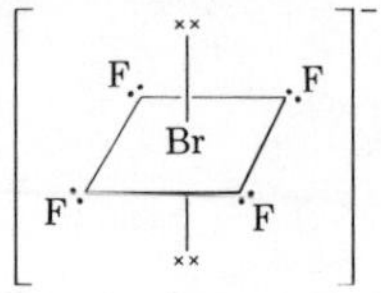

$$\left[\begin{array}{c} F\cdots\!\!\overset{\times\times}{\underset{Br}{|}}\!\!\cdots F \\ F\cdots\;\;\;\;\;\cdots F \\ \times\times \end{array} \right]^{-}$$

FIG. A.5 Structure of an interhalogen ion, BrF_4^-.

   (c) (1) square planar; (2) octahedral

18. Contrary conditions of stability prevail with the respective gaseous atoms and ions. In the neutral atom, the $3d$ orbitals are higher in energy than the $4s$ orbital; hence, the $4s$ fills first. In the gaseous ion, however, the $4s$ orbital is the higher in energy; hence, the electrons are preferentially retained in the lower-energy $3d$ orbitals and consequently, electrons migrate first from the higher-energy $4s$ orbital. Only when the $4s$ orbital is completely evacuated are $3d$ electrons lost and trivalent ions formed.

19. (a)

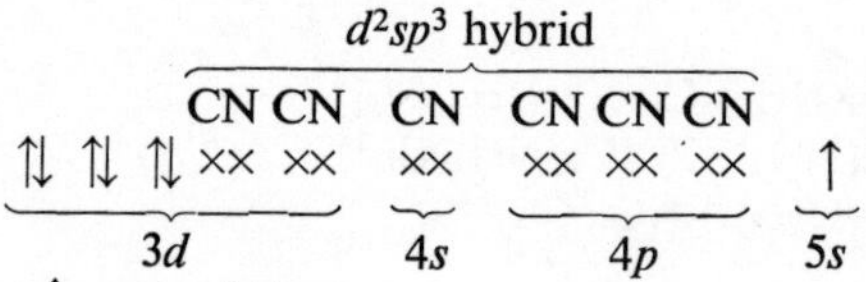

$$\underbrace{\overset{}{\uparrow\!\downarrow\;\uparrow\!\downarrow\;\uparrow\!\downarrow}\;\overset{CN\,CN}{\times\!\times\;\times\!\times}\;\overset{CN}{\times\!\times}\;\overset{CN\,CN\,CN}{\times\!\times\;\times\!\times\;\times\!\times}}_{\text{d^2sp^3 hybrid}}\;\underset{5s}{\uparrow}$$

$$\underbrace{}_{3d}\;\underbrace{}_{4s}\;\underbrace{}_{4p}\;\underbrace{}_{5s}$$

   (b) $\uparrow\;\;-\;\;(e_g)$
       $\uparrow\!\downarrow\;\uparrow\!\downarrow\;\uparrow\!\downarrow\;\;(t_{2g})$

20.

$$\left[\underbrace{\uparrow\!\downarrow\;\uparrow\!\downarrow\;\uparrow\!\downarrow}_{3d}\;\underbrace{\overset{H_2O\;H_2O\;H_2O\;H_2O\;H_2O\;H_2O}{\times\!\times\;\times\!\times\;\times\!\times\;\times\!\times\;\times\!\times\;\times\!\times}}_{\overbrace{4s\quad\;\;\;4p}^{d^2sp^3\text{ hybrid}}}\;\underset{5s}{\uparrow}\right] \to \left[\underbrace{\uparrow\!\downarrow\;\uparrow\!\downarrow\;\uparrow\;\uparrow\;\uparrow}_{3d}\;\underbrace{\overset{Cl\;Cl\;Cl\;Cl}{\times\!\times\!\times\!\times\;\times\!\times\;\times\!\times}}_{\overbrace{4s\;\;4p}^{sp^3\text{ hybrid}}}\right]$$

$Co(H_2O)_6^{2+}$ $\quad\quad\quad\quad\quad\quad\quad\quad\quad\quad$ $CoCl_4^{2-}$

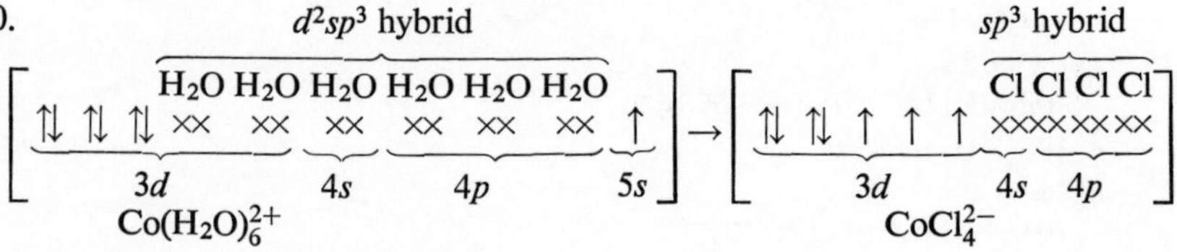

APPENDIX A 159

21. (a) $[Pt(NH_3)_2I_4]^0$
    (b) $[Pt(NH_3)_4Br_2]Br_2$
    (c) $K_2[Co(NH_3)(H_2O)Cl_4]$
    (d) $K[Co(NH_3)_2(H_2O)_2Cl_2]$
    (e) $[Ni(en)_2Cl_2]^{2+}$
    (f) $[Ni(H_2O)F_5]^-$
    (g) $[Cr(en)I_4]^-$
    (h) $([Fe(H_2O)_4(C_2O_4)])_2 SO_4$
    (i) $[Fe_3[FeF_6]_2$
    (j) $([Co(NH_3)_6][CrCl_6])^0$
    (k) $([Ag(NH_3)_2]_2[PtCl_4])^0$

CHAPTER 4. Aspects of MO Theory

1. Despite the net equality of calculated bond orders (each equivalent to a one-electron σ bond) the H_2^+ molecule-ion would be the more stable. This follows from the experimental revelation that an electron in an antibonding MO (the H_2^- has one such electron) is slightly more repulsing than an electron in a bonding orbital is attracting. As it lacks any antibonding electrons in its molecular valence-shell, the H_2^+ has a slight positive advantage over the H_2^- with respect to bonding stability, as expressing the magnitude of bond-order. That is, the H_2^- bond order is actually just a little bit less, numerically, than the net value of $\frac{1}{2}$ that has been calculated for it. A fairly reliable rule is that the greater the bond order, the shorter the bond length; and the greater the bond dissociation energy, the stronger the bonding. This is sustained by experimental evidence.

2. $C = KK\sigma_{1s}^2 \sigma_{1s}^{*2} (\sigma_{2p}^2 \pi_{2p}^1 \pi_{2p}^1)$
   yielding a bond order of $[(4 - 0/2] = 2$; a bond type equivalent to a net double bond constituted of two half-(single-electron) pi-bonds and one single sigma bond; and a paramagnetic response equivalent to the spins of two unpaired electrons.

3. (a) $O_2^0(= 16e^-) - 1e^- \to O_2^+(= 15e^-)$; hence
      $O_2^+ = KK\sigma_{2s}^2 \sigma_{2s}^{*2} (\pi_{2p}^2 \pi_{2p}^2 \sigma_{2p}^2 \pi_{2p}^{*1} \pi_{2p}^{*0})$
   (b) $N_2^0(= 14e^-) - 1e^- \to N_2^+(= 13e^-)$; hence
      $N_2^+ = KK\sigma_{2s}^2 \sigma_{2s}^{*2} (\pi_{2p}^2 \pi_{2p}^2 \sigma_{2p}^1)$

4. Bond order of $O_2^+ = [(6 - 1)/2] = 2.5$; one unpaired electron
   Bond order of $N_2^+ = [(5 - 0)/2] = 2.5$; one unpaired electron

5. In the O_2^+ molecule-ion, there is a valence-antibonding electron; in the N_2^+ species, there is none. Consequently, because the antibonding electron is slightly more effective in decreasing the strength of bonding than is the bonding electron in increasing the strength of bonding, the subtractive component in the computation of bond order of O_2^+ is a bit larger than actually employed. Therefore, the bond order of the O_2^+ species is just a bit less than the calculated net of 2.5. And its stability, consequently, is correspondingly smaller than that of the N_2^+ species.

6. (a) $Na_2 = KKLL(\sigma_{3s}^2)$
   $K_2 = KKLLMM(\sigma_{4s}^2)$
   $Rb_2 = KKLLMMNN(\sigma_{5s}^2)$
   $Cs_2 = KKLLMMNNOO(\sigma_{6s}^2)$
   Each has a bond order of $[(2-0)/2] = 1$, equivalent to a single σ bond.

   (b) $F_2 = KK\sigma_{2s}^2\sigma_{2s}^{*2}(\sigma_{2p}^2\pi_{2p}^2\pi_{2p}^2\pi_{2p}^{*2}\pi_{2p}^{*2})$
   $Cl_2 = KKLL\sigma_{3s}^2\sigma_{3s}^{*2}(\sigma_{3p}^2\pi_{3p}^2\pi_{3p}^2\pi_{3p}^{*2}\pi_{3p}^{*2})$
   $Br_2 = KKLLMM\sigma_{4s}^2\sigma_{4s}^{*2}(\sigma_{4p}^2\pi_{4p}^2\pi_{4p}^2\pi_{4p}^{*2}\pi_{4p}^{*2})$
   $I_2 = LLKKMMNN\sigma_{5s}^2\sigma_{5s}^{*2}(\sigma_{5p}^2\pi_{5p}^2\pi_{5p}^2\pi_{5p}^{*2}\pi_{5p}^{*2})$
   Each has a bond order of $[(6-4)/2] = 1$, equivalent to a net of a single σ bond — the attractions of the four bonding π electrons being cancelled by the repulsions of the four antibonding π electrons.

7. (a) 36 electrons; (b) 16 electrons; (c) 36 electrons; (d) 36 electrons; (e) 16 electrons.

8. $NO^+ = KK\sigma_{2s}^2\sigma_{2s}^{*2}(\sigma_{2p}^2\pi_{2p}^2\pi_{2p}^2)$
   Bond order $= [(6-0)/]2 = 3$, a triple bond: two π and one σ portrayable as
   $$[:N ::: O:]^+$$

9. Not possible. The order of filling orbitals, and the differentiations existing among bonding, antibonding, and nonbonding electrons, never permit the establishing of more than one sigma bond between any two atoms.

10. Energetically they are all degenerate but their spatial orientations are different. With the x Cartesian coordinate representing the internuclear axis of the atoms in the molecule, the one π orbital (or bond) must be oriented along the y coordinate (e.g., π_{2py}); whereas the other must be oriented along the z coordinate axis (e.g., π_{2pz}). Comparable interpretations prevail when considering either the y or the z coordinate to be the internuclear axis.

11. (a) free radical: $O_2 = KK\sigma_{2s}^2\sigma_{2s}^{*2}(\sigma_{2p}^2\pi_{2p}^2\pi_{2p}^2\pi_{2p}^{*1}\pi_{2p}^{*1})$.
    (b) diradical: $O_2 = KK\sigma_{2s}^2\sigma_{2s}^{*2}(\sigma_{2p}^2\pi_{2p}^2\pi_{2p}^2\pi_{2p}^{*1}\pi_{2p}^{*1})$.
    (c) not a free radical: $CO = KK\sigma_{2s}^2\sigma_{2s}^{*2}(\pi_{2p}^2\pi_{2p}^2\sigma_{2p}^2)$.

APPENDIX B: Energy Units, Fundamental Constants, and Reference Symbols

ENERGY

calorie: Amount of heat required to raise the temperature of one gram of pure water by one degree centigrade. As this quantity varies slightly for different one-degree intervals, depending upon the initial temperature of the water, the unit of the calorie is more accurately defined as the "15° calorie" rigorously representative of the interval 14.5-15.5 degree of centigrade temperature. A kilocalorie equals 1000 calories.

electron-volt: Kinetic energy gained by an electron when accelerated in an electrical field by a potential difference of one volt.

erg: The fundamental unit of energy in the metric or cgs (centimeter-gram-second) system. Its dimensions are gram cm^2 sec^{-2}.

Some Useful Conversions:

$$1 \text{ calorie (cal)} = 4.1840 \text{ joules}$$
$$1 \text{ electron-volt (eV)} = 1.6021 \times 10^{-12} \text{ erg}$$
$$1 \text{ erg} = 2.39 \times 10^{-11} \text{ kcal} = 2.39 \times 10^{-8} \text{ cal}$$
$$1 \text{ joule} = 10^7 \text{ ergs}$$

and, consequently,

$$1 \text{ eV/particle} = 1.6021 \times 10^{-12} \text{ erg/particle} = 23.060 \text{ kcal/mole}$$
$$1 \text{ erg/particle} = 6.2419 \times 10^{11} \text{ eV/particle} = 1.4396 \times 10^{13} \text{ kcal/mole}$$
$$1 \text{ kcal/mole} = 6.9468 \times 10^{-14} \text{ erg/particle} = 4.3361 \times 10^{-2} \text{ eV/particle}$$

CONSTANTS

Symbol	Description	Value
m_e	Mass of electron, at rest	5.486×10^{-4} amu
		9.107×10^{-28} gram
e	Charge upon electron	4.80298×10^{-10} esu
		1.6021×10^{-19} coulomb
m_p	Mass of proton, at rest	1.0073 amu
		1.6724×10^{-24} gram
p	Charge upon proton	1.6021×10^{-19} coulomb
N	Avogadro Number	6.02296×10^{23} molecules/mole
h	Planck Constant	6.6256×10^{-27} erg sec/particle
f	Faraday Constant	96,489 coulombs/equivalent
		23,060 calories/volt-equivalent
c	Speed of light, *in vacuo*	2.99793×10^{10} cm/sec
k	Boltzmann Constant	1.3805×10^{-16} erg/°K molecule

APPENDIX C: International Atomic Weights

Atomic weights are based on the isotope of carbon of mass number 12.0000... Values in parentheses, assigned to radioactive elements, denote mass numbers of the most stable isotopes.

APPENDIX C

Name	Symbol	Atomic Number	Atomic Weight
Actinium	Ac	89	(227)
Aluminum	Al	13	26.9815
Americium	Am	95	(243)
Antimony	Sb	51	121.75
Argon	Ar	18	39.948
Arsenic	As	33	74.9216
Astatine	At	85	(210)
Barium	Ba	56	137.34
Berkelium	Bk	97	(249)
Beryllium	Be	4	9.0122
Bismuth	Bi	83	208.920
Boron	B	5	10.811
Bromine	Br	35	79.909
Cadmium	Cd	48	112.40
Calcium	Ca	20	40.08
Californium	Cf	98	(251)
Carbon	C	6	12.01115
Cerium	Ce	58	140.12
Cesium	Cs	55	132.905
Chlorine	Cl	17	35.453
Chromium	Cr	24	51.996
Cobalt	Co	27	58.9332
Copper	Cu	29	63.54
Curium	Cm	96	(245)
Dysprosium	Dy	66	162.50
Einsteinium	Es	99	(254)
Erbium	Er	68	167.26
Europium	Eu	63	151.96
Fermium	Fm	100	(253)
Fluroine	F	9	18.9984
Francium	Fr	87	(223)
Gadolinium	Gd	64	157.25
Gallium	Ga	31	69.72
Germanium	Ge	32	72.59
Gold	Au	79	196.967
Hafnium	Hf	72	178.49
Helium	He	2	4.0026
Holmium	Ho	67	164.930
Hydrogen	H	1	1.00797
Indium	In	49	114.82
Iodine	I	53	126.9044
Iridium	Ir	77	192.2
Iron	Fe	26	55.847
Krypton	Kr	36	83.80
Lanthanum	La	57	138.91
Lawrencium	Lw	103	(257)
Lead	Pb	82	207.19
Lithium	Li	3	6.939
Lutetium	Lu	71	174.97

APPENDIX C

Name	Symbol	Atomic Number	Atomic Weight
Magnesium	Mg	12	24.312
Manganese	Mn	25	54.9381
Mendelevium	Md	101	(256)
Mercury	Hg	80	200.59
Molybdenum	Mo	42	95.94
Neodymium	Nd	60	144.24
Neon	Ne	10	20.183
Neptunium	Np	93	(237)
Nickel	Ni	28	58.71
Niobium	Nb	41	92.906
Nitrogen	N	7	14.0067
Nobelium	No	102	(254)
Osmium	Os	76	190.2
Oxygen	O	8	15.9994
Palladium	Pd	46	106.4
Phosphorus	P	15	30.9738
Platinum	Pt	78	195.09
Plutonium	Pu	94	(242)
Polonium	Po	84	(210)
Potassium	K	19	39.102
Praseodymium	Pr	59	140.907
Promethium	Pm	61	(147)
Protoactinium	Pa	91	(231)
Radium	Ra	88	(226)
Radon	Rn	86	(222)
Rhenium	Re	75	186.2
Rhodium	Rh	45	102.905
Rubidium	Rb	37	85.47
Ruthenium	Ru	44	101.07
Samarium	Sm	62	150.35
Scandium	Sc	21	44.956
Selenium	Se	34	78.96
Silicon	Si	14	28.086
Silver	Ag	47	107.870
Sodium	Na	11	22.9898
Strontium	Sr	38	87.62
Sulfur	S	16	32.064
Tantalum	Ta	73	180.948
Technetium	Tc	43	(99)
Tellurium	Te	52	127.60
Terbium	Tb	65	158.924
Thallium	Tl	81	204.37
Thorium	Th	90	232.038
Thulium	Tm	69	168.934
Tin	Sn	50	118.69
Titanium	Ti	22	47.90
Tungsten	W	74	183.85
Uranium	U	92	238.03
Vanadium	V	23	50.942

Name	Symbol	Atomic Number	Atomic Weight
Xenon	Xe	54	131.30
Ytterbium	Yb	70	173.04
Yttrium	Y	39	88.905
Zinc	Zn	30	65.37
Zirconium	Zr	40	91.22

GREEK ALPHABET

Many of these letters are employed in the text to symbolize various physicochemical terms and concepts. They should be familiarly associated with their corresponding word names whenever encountered.

alpha	A	α
beta	B	β
gamma	Γ	γ
delta	Δ	δ, ∂
epsilon	E	ϵ
zeta	Z	ζ
eta	H	η
theta	Θ	θ, ϑ
iota	I	ι
kappa	K	κ
lambda	Λ	λ
mu	M	μ
nu	N	ν
xi	Ξ	ξ
omicron	O	o
pi	Π	π
rho	P	ρ
sigma	Σ	σs
tau	T	τ
upsilon	Υ	υ
phi	Φ	ϕ, φ
chi	X	χ
psi	Ψ	ψ
omega	Ω	ω

APPENDIX D

Periodicity of Properties of the Elements

KEY
- (a) first ionization potential, in electron volts
- (b) electronegativity in arbitrary units: Pauling scale, *left;* Mulliken scale, *right*
- (c) atomic radius in covalent bondings, in Angstrom units
- (d) ionic radius of specified species, in crystals thereof, in Angstrom units

	Main Group I	Main Group II
1s fills	**1 H** $1s^1$ (a) 13.60 (b) 2.1; 2.1 (c) 0.37 (d) 0.29 H^+	
2s fills	**3 Li** [He] $2s^1$ (a) 5.39 (b) 1.0; 0.97 (c) 1.23 (d) 0.60 Li^+	**4 Be** [He] $2s^2$ (a) 9.32 (b) 1.5; 1.47 (c) 0.89 (d) 0.30 Be^{2+}
3s fills	**11 Na** [Ne] $3s^1$ (a) 5.14 (b) 0.9; 1.01 (c) 1.57 (d) 0.95 Na^+	**12 Mg** [Ne] $3s^2$ (a) 7.64 (b) 1.2; 1.23 (c) 1.36 (d) 0.66 Mg^{2+}
4s fills	**19 K** [Ar] $4s^1$ (a) 4.34 (b) 0.8; 0.91 (c) 2.03 (d) 1.33 K^+	**20 Ca** [Ar] $4s^2$ (a) 6.11 (b) 1.0; 1.04 (c) 1.74 (d) 0.99 Ca^{2+}
5s fills	**37 Rb** [Kr] $5s^1$ (a) 4.18 (b) 0.8; 0.89 (c) 2.16 (d) 1.48 Rb^+	**38 Sr** [Kr] $5s^2$ (a) 5.69 (b) 1.0; 0.99 (c) 1.91 (d) 1.15 Sr^{2+}
6s fills	**55 Cs** [Xe] $6s^1$ (a) 3.89 (b) 0.7; 0.86 (c) 2.35 (d) 1.67 Cs^+	**56 Ba** [Xe] $6s^2$ (a) 5.21 (b) 0.9; 0.97 (c) 1.98 (d) 1.37 Ba^{2+}
7s fills	**87 Fr** [Rn] $7s^1$ (a) — (b) 0.7; 0.86 (c) — (d) 1.75 Fr^+	**88 Ra** [Rn] $7s^2$ (a) 5.28 (b) 0.9; 0.97 (c) — (d) 1.55 Ra^{2+}

APPENDIX D

	Transition Group III	Transition Group IV	Transition Group V	Transition Group VI
$3d$ fills	**21 Sc** [Ar] $3d^1 4s^2$ (a) 6.56 (b) 1.3; 1.20 (c) 1.44 (d) 0.81 Sc^{3+}	**22 Ti** [Ar] $3d^2 4s^2$ (a) 6.83 (b) 1.5; 1.32 (c) 1.32 (d) 0.68 Ti^{4+}	**23 V** [Ar] $3d^3 4s^2$ (a) 6.74 (b) 1.6; 1.45 (c) 1.22 (d) 0.69 V^{3+}	**24 Cr** [Ar] $3d^5 4s^1$ (a) 6.76 (b) 1.6; 1.56 (c) 1.17 (d) 0.62 Cr^{3+}
$4d$ fills	**39 Y** [Kr] $4d^1 5s^2$ (a) 6.38 (b) 1.2; 1.11 (c) 1.62 (d) 0.96 Y^{3+}	**40 Zr** [Kr] $4d^2 5s^2$ (a) 6.84 (b) 1.4; 1.22 (c) 1.45 (d) 0.87 Zr^{4+}	**41 Nb** [Kr] $4d^4 5s^1$ (a) 6.88 (b) 1.6; 1.23 (c) 1.34 (d) 0.78 Nb^{3+}	**42 Mo** [Kr] $4d^5 5s^1$ (a) 7.13 (b) 1.8; 1.30 (c) 1.29 (d) 0.62 Mo^{6+}
$5d$ fills	**57 La*** [Xe] $5d^1 6s^2$ (a) 5.61 (b) 1.1; 1.08 (c) 1.69 (d) 1.16 La^{3+}	**72 Hf** [Xe] $4f^{14} 5d^2 6s^2$ (a) 5.5 (b) 1.3; 1.23 (c) 1.44 (d) 0.86 Hf^{4+}	**73 Ta** [Xe] $4f^{14} 5d^3 6s^2$ (a) 7.7 (b) 1.5; 1.33 (c) 1.34 (d) 0.77 Ta^{3+}	**74 W** [Xe] $4f^{14} 5d^4 6s^2$ (a) 7.98 (b) 1.7; 1.40 (c) 1.30 (d) 0.68 W^{6+}
$6d$ fills	**89 Ac**** [Rn] $6d^1 7s^2$ (a) — (b) 1.1; 1.00 (c) — (d) 1.11 Ac^{3+}	*LANTHANIDES* * $4f$ 58-71 fills *ACTINIDES* ** $5f$ 90-103 fills	**58 Ce** [Xe] $4f^2 5d^0 6s^2$ **90 Th** [Rn] $5f^0 6d^2 7s^2$	**59 Pr** [Xe] $4f^3 5d^0 6s^2$ **91 Pa** [Rn] $5f^2 6d^1 7s^2$

APPENDIX D

	Transition Group VII	Transition Group VII		
3d fills	**25 Mn** [Ar] $3d^5 4s^2$ (a) 7.43 (b) 1.5; 1.60 (c) 1.17 (d) 0.78 Mn^{2+}	**26 Fe** [Ar] $3d^6 4s^2$ (a) 7.90 (b) 1.8; 1.64 (c) 1.17 (d) 0.76 Fe^{2+}; 0.64 Fe^{3+}	**27 Co** [Ar] $3d^7 4s^2$ (a) 7.86 (b) 1.8; 1.70 (c) 1.16 (d) 0.74 Co^{2+}	**28 Ni** [Ar] $3d^8 4s^2$ (a) 7.63 (b) 1.8; 1.75 (c) 1.15 (d) 0.73 Ni^{2+}
4d fills	**43 Tc** [Kr] $4d^6 5s^1$ (a) 7.23 (b) 1.9; 1.36 (c) — (d) —	**44 Ru** [Kr] $4d^7 5s^1$ (a) 7.36 (b) 2.2; 1.42 (c) 1.24 (d) 0.64 Ru^{4+}	**45 Rh** [Kr] $4d^8 5s^1$ (a) 7.46 (b) 2.2; 1.45 (c) 1.25 (d) 0.69 Rh^{3+}	**46 Pd** [Kr] $4d^{10} 5s^0$ (a) 8.33 (b) 2.2; 1.35 (c) 1.28 (d) 0.50 Pd^{2+}
5d fills	**75 Re** [Xe] $4f^{14} 5d^5 6s^2$ (a) 7.87 (b) 1.9; 1.46 (c) 1.28 (d) 0.86 Re^{2+}	**76 Os** [Xe] $4f^{14} 5d^6 6s^2$ (a) 8.7 (b) 2.2; 1.52 (c) 1.26 (d) 0.67 Os^{4+}	**77 Ir** [Xe] $4f^{14} 5d^7 6s^2$ (a) 9.2 (b) 2.2; 1.55 (c) 1.26 (d) 0.66 Ir^{4+}	**78 Pt** [Xe] $4f^{14} 5d^9 6s^1$ (a) 9.0 (b) 2.2; 1.44 (c) 1.29 (d) 0.52 Pt^{2+}
4f fills	**60 Nd** [Xe] $4f^4 5d^0 6s^2$	**61 Pm** [Xe] $4f^5 5d^0 6s^2$	**62 Sm** [Xe] $4f^6 5d^0 6s^2$	**63 Eu** [Xe] $4f^7 5d^0 6s^2$
5f fills	**92 U** [Rn] $5f^3 6d^1 7s^2$	**93 Np** [Rn] $5f^4 6d^1 7s^2$	**94 Pu** [Rn] $5f^5 6d^1 7s^2$	**95 Am** [Rn] $5f^6 6d^1 7s^2$

APPENDIX D

				Main Group III	Main Group IV
			2p fills	**5 B** [He] $2s^2 2p^1$ (a) 8.30 (b) 2.0; 2.01 (c) 0.80 (d) —	**6 C** [He] $2s^2 2p^2$ (a) 11.26 (b) 2.5; 2.50 (c) 0.77 (d) 2.60 C^{4-}
			3p fills	**13 Al** [Ne] $3s^2 3p^1$ (a) 5.98 (b) 1.5; 1.74 (c) 1.25 (d) 0.52 Al^{3+}	**14 Si** [Ne] $3s^2 3p^2$ (a) 8.15 (b) 1.8; 1.47 (c) 1.17 (d) 2.71 Si^{4-}
	Post Transition Group I	Post Transition Group II			
	29 Cu [Ar] $3d^{10} 4s^1$ (a) 7.72 (b) 1.9; 1.75 (c) 1.17 (d) 0.93 Cu^+; 0.69 Cu^{2+}	**30 Zn** [Ar] $3d^{10} 4s^2$ (a) 9.39 (b) 1.6; 1.66 (c) 1.25 (d) 0.72 Zn^{2+}	4p fills	**31 Ga** [Ar] $3d^{10} 4s^2 4p^1$ (a) 6.00 (b) 1.6; 1.82 (c) 1.25 (d) 0.60 Ga^{3+}	**32 Ge** [Ar] $3d^{10} 4s^2 4p^2$ (a) 7.88 (b) 1.8; 2.02 (c) 1.22 (d) 0.53 Ge^{4+}
	47 Ag [Kr] $4d^{10} 5s^1$ (a) 7.57 (b) 1.9; 1.42 (c) 1.34 (d) 1.26 Ag^+	**48 Cd** [Kr] $4d^{10} 5s^2$ (a) 8.99 (b) 1.7; 1.46 (c) 1.41 (d) 0.96 Cd^{2+}	5p fills	**49 In** [Kr] $4d^{10} 5s^2 5p^1$ (a) 5.78 (b) 1.7; 1.49 (c) 1.50 (d) 0.81 In^{3+}	**50 Sn** [Kr] $4d^{10} 5s^2 5p^2$ (a) 7.33 (b) 1.8; 1.72 (c) 1.41 (d) 1.10 Sn^{2+}
	79 Au [Xe] $4f^{14} 5d^{10} 6s^1$ (a) 9.22 (b) 2.4; 1.42 (c) 1.34 (d) 1.37 Au^+	**80 Hg** [Xe] $4f^{14} 5d^{10} 6s^2$ (a) 10.43 (b) 1.9; 1.44 (c) 1.44 (d) 1.10 Hg^{2+}	6p fills	**81 Tl** [Xe] $4f^{14} 5d^{10} 6s^2 6p^1$ (a) 6.11 (b) 1.8; 1.44 (c) 1.55 (d) 0.95 Tl^{3+}	**82 Pb** [Xe] $4f^{14} 5d^{10} 6s^2 6p^2$ (a) 7.42 (b) 1.8; 1.55 (c) 1.54 (d) 1.27 Pb^{2+}

64 Gd [Xe] $4f^7 5d^1 6s^2$	**65 Tb** [Xe] $4f^9 5d^0 6s^2$		**66 Dy** [Xe] $4f^{10} 5d^0 6s^2$	**67 Ho** [Xe] $4f^{11} 5d^0 6s^2$
96 Cm [Rn] $5f^7 6d^1 7s^2$	**97 Bk** [Rn] $5f^8 6d^1 7s^2$		**98 Cf** [Rn] $5f^9 6d^1 7s^2$	**99 Es** [Rn] $5f^{10} 6d^1 7s^2$

APPENDIX D

	Main Group V	Main Group VI	Main Group VII	Main Group 0
				2 He $1s^2$ (a) 24.58 (b) — (c) — (d) —
2p fills	**7 N** [He] $2s^2 2p^3$ (a) 14.54 (b) 3.0; 3.07 (c) 0.74 (d) 1.71 N^{3-}	**8 O** [He] $2s^2 2p^4$ (a) 13.61 (b) 3.5; 3.50 (c) 0.74 (d) 1.40 O^{2-}	**9 F** [He] $2s^2 2p^5$ (a) 17.42 (b) 4.0; 4.10 (c) 0.72 (d) 1.34 F^-	**10 Ne** [He] $2s^2 2p^6$ (a) 21.56 (b) — (c) — (d) —
3p fills	**15 P** [Ne] $3s^2 3p^3$ (a) 10.55 (b) 2.1; 2.06 (c) 1.10 (d) 2.12 P^{3-}	**16 S** [Ne] $3s^2 3p^4$ (a) 10.36 (b) 2.5; 2.44 (c) 1.04 (d) 1.84 S^{2-}	**17 Cl** [Ne] $3s^2 3p^5$ (a) 13.01 (b) 3.0; 2.83 (c) 0.99 (d) 1.81 Cl^-	**18 Ar** [He] $3s^2 3p^6$ (a) 15.76 (b) — (c) — (d) —
4p fills	**33 As** [Ar] $3d^{10} 4s^2 4p^3$ (a) 9.81 (b) 2.0; 2.20 (c) 1.21 (d) 2.22 As^{3-}	**34 Se** [Ar] $3d^{10} 4s^2 4p^4$ (a) 9.75 (b) 2.4; 2.48 (c) 1.17 (d) 1.98 Se^{2-}	**35 Br** [Ar] $3d^{10} 4s^2 4p^5$ (a) 11.84 (b) 2.8; 2.74 (c) 1.14 (d) 1.96 Br^-	**36 Kr** [Ar] $3d^{10} 4s^2 4p^6$ (a) 14.00 (b) — (c) — (d) —
5p fills	**51 Sb** [Kr] $4d^{10} 5s^2 5p^3$ (a) 8.64 (b) 1.9; 1.82 (c) 1.41 (d) 0.92 Sb^{3+}	**52 Te** [Kr] $4d^{10} 5s^2 5p^4$ (a) 9.01 (b) 2.1; 2.01 (c) 1.37 (d) 2.21 Te^{2-}	**53 I** [Kr] $4d^{10} 5s^2 5p^5$ (a) 10.44 (b) 2.5; 2.29 (c) 1.33 (d) 2.19 I^-	**54 Xe** [Kr] $4d^{10} 5s^2 5p^6$ (a) 12.12 (b) — (c) — (d) —
6p fills	**83 Bi** [Xe] $4f^{14} 5d^{10} 6s^2 6p^3$ (a) 7.29 (b) 1.9; 1.67 (c) 1.52 (d) 1.20 Bi^{3+}	**84 Po** [Xe] $4f^{14} 5d^{10} 6s^2 6p^4$ (a) 8.43 (b) 2.0; 1.76 (c) 1.53 (d) —	**85 At** [Xe] $4f^{14} 5d^{10} 6s^2 6p^5$ (a) — (b) 2.2; 1.90 (c) — (d) —	**86 Rn** [Xe] $4f^{14} 5d^{10} 6s^2 6p^6$ (a) 10.74 (b) — (c) — (d) —
4f fills	**68 Er** [Xe] $4f^{12} 5d^0 6s^2$	**69 Tm** [Xe] $4f^{13} 5d^0 6s^2$	**70 Yb** [Xe] $4f^{14} 5d^0 6s^2$	**71 Lu** [Xe] $4f^{14} 5d^1 6s^2$
5f fills	**100 Fm** [Rn] $5f^{11} 6d^1 7s^2$	**101 Md** [Rn] $5f^{12} 6d^1 7s^2$	**102 No** [Rn] $5f^{13} 6d^1 7s^2$	**103 Lw** [Rn] $5f^{14} 6d^1 7s^2$

INDEX

Antibonding MO, 126, 128, 129–144
Atom, shell arrangement, 12
Atomic number, 5, 11–12
   spectra and, 11–12
Atomic orbitals, 18–21
   arrangements of, 19
   ground states of, 20–21
   LCAO, 130–132
   in MO formation, 125–144
   shapes of, 18–20
Atomic properties, 1–27
Atomic spectra, origins, 11–12
Atomic weights/numbers, table, 162–164
Aufbau, 132–134
Azimuthal quantum number, 16

Balmer spectra, 24
Band theory, metals, 69–72
Bethe, CF theory, 83

Bohr, 14
Bohr hydrogen model, 22
   line emission spectra of, 24
Boltzmann, 45
Bond angle, 48–51
Bond direction, 51–57
   linear, 51–52
   octahedral, 56–57
   tetrahedral, 53–54
   triangular planar, 52–53
   trigonal bipyramidal, 54–56
Bond dissociation, energy, 63
Bond energy, 9, 33–36
   in Born-Haber cycle, 63
Bond length, 36–39, 50
Bond order, 38–39, 135
Bond polarity, 29–32
Bond stability, charge delocalization, 131–132
Bond strength, 32–36
Bonding electron pairs, 48
Bonding MO, 126, 128, 129–144

INDEX

Born-Haber cycle, 62–66
Brackett spectra, 24
Bridge isomerism, 121–122

Capacitance, 42
Carbonyls, transition metals, 119
Chelate/chelation, 85–86
Chemical bond, 29–78
  electron cloud structures of, 57–61
  ionic character of, 29–32
  polarity of, 29–32
Chemical bonding, causes, 1–4
Cis-trans isomerism, 91–92, 120
Color, transition complexes, 115–116
Complex ions and MO, 139–141
Complexes, 89–124
  colors of, 115–116
  inner-orbital, 107–108
  lability of, 97–98
  linear 2-coordination, 90
  nomenclature of, 94–96
  octahedral 6-coordination, 90–94, 143
  outer-orbital, 108–109
  spatial orientations of, 89–94, 139–141
  square planar 4-coordination, 90
  stability of, 97–98
  tetrahedral 4-coordination, 90
Complexing, transitional, 79–123, 139–141
  CF interpretations of, 83
  LF interpretations of, 83
  MO interpretations of, 83, 139–141
  VB interpretations of, 83–84
Conductor, metal, 69–71
Constants, numerical values, 161
Coordination isomerism, 121
Coordination number, 85
Coordination sphere, 87–94
  experimental basis for, 87–88
Coulomb, electrostatics law, 44
  dielectric constant from, 44
Covalent bond, 3–4
Crystal fields, 98–114
  distortions in, 105–106
  octahedral, 101–111

Crystal fields (cont'd)
  splitting of, 99–101
  square planar, 111–114
  strengths of, 100–101
  tetrahedral, 111–114
  theory of, 83
  valence bonds vs., 98–99, 106–111, 114–115

d orbitals, 18–20
dsp^2 hybridization, 110, 111
d^2sp^3 hybridization, 107, 108
de Broglie, quantum mechanics, 14–15, 27
Debye, 41
Degenerate MO, 126, 128, 141–142
Delta bonds, 58–61
Diamagnetism, 17
Dielectric constant, 41–42
  derivation from Coulomb's law, 44
  derivation of dipole moment from, 44–46
Dipole moment, 41–46
  derivation from dielectric constant, 44–46
  induced, 42, 45
Diradical, 147
Distortions, crystal fields, 105–106

Effective nuclear charge, 5–6
Einstein, quantum theory, 7
Electron affinity, 8
  in Born-Haber cycle, 63
Electron binding and nuclear charge, 26
Electron energy, 13–14, 22
Electron ground state, 13
Electron location, probability, 14
Electron, momentum, 14
Electron transitions, spectra, 11–12
Electronegativity, 8–11, 30–31
  numerical assignments of, 9–11
  sequence in halogens, 30–31
Electrons, energies/distributions, 13–16
Energy, molecular orbitals, 125–129

INDEX

Energy quantum of electron, 7
Energy units, 161
Energy-stability, octahedral complexes, 143
Expanded octets, theory, 73–74

f orbitals, 20
Formal charge, 75–77
Free electron theory, metals, 68–69
Free radical, 147

Geometric symmetry, 89–94
Gouy balance, 17
Greek alphabet, 164
Ground-state filling, AO, 20–21
Ground states, MO, 127, 136–139
  heteronuclear diatomic, 136–139
  homonuclear diatomic, 138–139

Half bonds, MO theory, 135
Head-to-head overlap, 58–59
Heat of formation, 63–66
Heat of reaction, bond energy, 33–35
Heisenberg, uncertainty principle, 14
Hess, law, 63
Homonuclear nonpolarity, 11
Hund, maximum multiplicity rule, 17
Hybridization, 40, 107–111
  dsp^2, 110, 111
  d^2sp^3, 107, 108
  sp, 111
  sp^3, 110, 111
  sp^3d^2, 108, 109
Hydrogen bonding, 46–47

Impurity semiconductor, 71
In phase, MO, 132–133
Inner-orbital complex, 107, 108
Insulator atoms, 69, 71–72

Ionic bond, 3–4
  bond polarity and, 29–32
Ionization energy, 4–6
  measurement of, 6–8
  in Born-Haber cycle, 63
Ionization isomerism, 120–121
Ionization potential, 4–6
  critical, 6
  measurement of, 6–8
Isoelectronic species, nuclear charge, 26
Isomerism, 91, 120–122
  bridge, 121–122
  cis-trans, 91–92, 120
  coordination, 121
  ionization, 120–121
  mode-of-attachment, 121
  multiform-empiric, 121
  optical, 91, 120

K-capture, 13

L-capture, 13
Lability, complexes, 97–98
Lattice energy, 63
Lattice structure, ionic, 61–62
Ligancy/ligands, 85
Ligand field theory, 83
Line emission spectra, 13, 24
Linear bond orientations, 51–52
Linear combination of atomic orbitals, 130–132
Linear molecules, 43
Lone electron pairs, 48
Lyman spectra, 24

Madelung, constant, 63
Magnetic moment, 80, 106–113
Magnetic quantum number, 16–17
Metal, "true", 70
Metallic bond, 66–72
  and free electron theory, 68–69
  and valence band theory, 69–72

INDEX

Mode-of-attachment isomerism, 121
Molecular orbital theory, crystal fields, 139–141
Molecular orbitals, 2–3, 125–145
  antibonding, 126, 128, 129–144
  bonding, 126, 128, 129–144
  complex ions and, 139–141
  delta, 126; 128
  degenerate, 126, 128
  doubly degenerate, 142
  half bonds in, 135
  hydrogen formation and, 2
  in phase, 132–133
  LCAO in, 130–132
  nonbonding, 126, 128, 129–144
  out of phase, 132–133
  pi, 126, 128
  sigma, 126, 128
  single-electron bonds in, 135
  singly degenerate, 141
  triply degenerate, 142
  wave function of, 130–132
Molecular shapes, 43
Moseley, law, 12
Mulliken, electronegativity, 10–11
Multiform-empiric isomerism, 121

Nomenclature, complexes, 94–96
Nonbonding MO, 126, 128, 129–144
Nuclear charge, effective, 5
Nuclear charge, electron binding, 26
Nuclear stability, 13

Octahedral bond orientations, 56–57
Octahedral complexes, 99–101, 141–144
  energy separations in, 141–144
  energy-stability of, 143
Octahedral crystal fields, 99–111
  splitting of, 99–101
Octet theory, expanded, 73–74
Optical isomerism, 91, 120
Orbital quantum number, 16
Out-of-phase, MO, 132–133
Outer-orbital complex, 108, 109
Overlap in MO, 133, 144–145

p orbitals, 18–19
Paramagnetism, 17, 144–145
  in strong overlap MO, 145
  in weak overlap MO, 145
Pauli, exclusion principle, 17
Pauling, 9
  elecronegativity scale, 9
  and VB theory, 83–84
Percent ionic character, 30–32
Pfund spectra, 24
Pi bonds, 58–61
Planck, constant, 7
Polarizability, 42
Post-transitional element, 82
Primary (electro-) valence, 87
Principal quantum number, 16

Quantum numbers, 16–17
  magnetic, 16
  orbital (azimuthal), 16
  principal, 16
  spin, 17
Quantum states, radiation energy, 22
  continuum of, 23

Radiation quanta, 22–23
Radius ratio, 86, 116–118
Resonance, 39–40
Ritz-Paschen spectra, 24
Rydberg constant, 23

s orbitals, 18–19
sp hybridization
sp^3 hybridization, 110, 111
sp^3d^2 hybridization, 108, 109
Schrödinger wave equation, 15–16
Screening (shielding) effect, 5
Secondary (auxillary) valence, 87
Semiconductor atoms, 70–71
Shielding (screening) effect, 5
Side-to-side overlap, 58–59
Sigma bonds, 58–61

Single-electron bonds, MO, 135
Slater, VB theory, 83
Spectra and atomic number, 11–12
Spectral continuum, 23
Spin quantum number, 17
Splitting, crystal fields, 99–101
Square planar crystal fields, 111–114
Stability, complexes, 97–98
Stability, molecular orbitals, 125–129
Stability, nuclear, 13
Stability of bond, charge delocalization, 131–132
Stereoisomerism, 89–94
Strong crystal field, 100
Strong overlap, MO, 144–145
Sublimation energy, 63
Symmetry distortions, 60

Tetrahedral bond orientations, 53–54
Tetrahedral crystal fields, 111–114
Thermodynamic cycle, 61–62
    Born-Haber, 62–66
Trans-cis isomerism, 91–92, 120
Transition metal complexing, 79–116
Transitional element, 82
Triangular planar bond orientations, 52–53

Triangular planar molecules, 43
Trigonal bipyramidal bond orientations, 54–56
Trigonal pyramidal molecules, 43

Valence and formal charge, 75–77
Valence band, metal, 69–72
Valence bond, theory, 83, 98–111
    crystal field vs., 98–99, 106–111, 114–115
Valence, primary, 87
Valence, secondary (auxillary), 87
van Vleck, MO theory, 83

Wavelength, 7
Wave frequency, 7
Wave function, 15, 130–132
Wave number, 23
Weak crystal field, 100
Weak overlap, MO, 144–145
Werner, 87

Zeeman effect, 16